Framing Farming:
Communication Strategies for Animal Rights

Critical Animal Studies 2

Framing Farming:
Communication Strategies for Animal Rights

Carrie P. Freeman

Rodopi

Amsterdam - New York, NY 2014

Critical Animal Studies

2. Carrie P. Freeman, *Framing Farming: Communication Strategies for Animal Rights.*
1. Kim Socha, *Women, Destruction, and the Avant-Garde. A Paradigm for Animal Liberation.*

This book is printed on recycled paper.

Cover photo: Jo-Anne McArthur / We Animals

The paper on which this book is printed meets the requirements of "ISO 9706:1994, Information and documentation - Paper for documents - Requirements for permanence".

ISBN: 978-90-420-3892-9
E-Book ISBN: 978-94-012-1174-1

Printed in The Netherlands

Printed by Printforce, the Netherlands

Table of Contents

PART II
HOW U.S. ANIMAL RIGHTS ORGANIZATIONS FRAME FOOD CAMPAIGN MESSAGES

List of Images Used in the Book

Foreword

Author's Background & Perspective

This book reflects my own American perspective as a long time vegan, animal activist, communication practitioner, and now communications professor. I first became involved in animal rights and environmental issues in 1989 as a freshman at the University of Florida, thanks to a vegan girl who lived on my dorm floor. As my college meal plan wasn't amenable, I vowed to go vegetarian upon graduation. By 1996 I had gradually shifted to a vegan diet, as I was especially encouraged to avoid eggs, dairy, and leather in preparation for a week long gig volunteering at PETA, where they expect interns to lead a vegan lifestyle. While working paid positions in public relations and professional development training, I founded and ran the grassroots group The Vegetarian Society of Southwest Florida for four years in the late 1990's. My interest in becoming a communication professor was spawned by a desire as an activist (hosting literature tables and documentaries) to uncover the methods and messages that could encourage other people to care about fellow animals as I did. So I went to the University of Georgia to earn my master's in media studies, where I co-founded the student animal rights group Speak Out for Species in 2003, writing my thesis on national news coverage of farmed animals. For my doctorate in communication, I headed to the University of Oregon in the eco-conscious Pacific Northwest, where I learned about forest protection and also helped to reinvigorate their Students for the Ethical Treatment of Animals group, serving as its co-director for a year while writing the dissertation that would become the foundation for this book.

Today, in my sixth year as an Assistant Professor of Communication at Georgia State University in Atlanta, I get to teach classes about media ethics, strategic communication for social change, and communicating about animal and environmental issues. Now that my paid job involves some animal advocacy, I'm taking a break from running grassroots groups. But I enjoy co-hosting a weekly animal rights radio show (Second Opinion Radio) with Melody Paris and Sonia Swartz and a semi-monthly environmental show (In Tune to Nature) on Atlanta's non-commercial indie station, WRFG (Radio Free Georgia), to give a much needed voice to activists who work on behalf of animals and nature.

Because of the magnitude of animal exploitation in the food industry and the myriad social and environmental benefits to which veganism contributes, I personally believe that promoting a plant-based diet should be a priority for the animal protection movement, as well as movements on behalf of human justice and the natural world. Therefore, I have made animal agribusiness and food a priority in my own activism as well as in my academic research. So this book represents the culmination of several decades of my own experience, reflection, and study on the subject of vegetarian advocacy.

Before reading the detailed review of scholarly literature and my analysis of vegan activism, readers may appreciate having a preview of the theories and worldviews that inform this study. I consider myself part of Leftist progressive politics, but I concentrate less on legal and political concepts than I do on ethical concepts. Ethical principles are ideally supposed to form the basis of cultural values that shape laws and public policy. Within ethics, I embrace respect for cultural diversity to a great extent, but I am less of a moral relativist and more of a moral absolutist, especially when it comes to larger issues of respecting life, equality, and freedoms for human and nonhuman animals. I don't think tradition or religion is an excuse for causing unnecessary harm to any sentient being. This respect for universal principles lends itself well to duty-based/deontological ethics as well as virtue ethics (although I talk more about "values" than virtues). Much of this book is about how we can design communication campaigns to character-build around values such as fairness, care and compassion, altruism, social responsibility, sustainability, non-violence, authenticity, and integrity.

Principles drawn from deep ecology and analytical animal rights philosophies extend my support for human social justice to foster broader respect for the lives of *all* animals and the natural world upon which we all depend. This relies on subverting hierarchical relationships related to dichotomies separating male/female, white/black, straight/gay, culture/nature, and human/animal. If we humans start to embrace and respect our own animality, I believe we would reduce our discrimination of animals (including other humans). To me, animal rights is a universal ethic that stands for anti-discrimination, anti-instrumentalism (against using anyone as a means to an end), and, thus, anti-domestication. My version of justice for nonhuman animals is allowing them the dignity to live free of human control or exploitation as a resource, which requires us to more equitably share the planet and preserve healthy habitats. I believe every sentient being has the right to life, liberty, and the pursuit of happiness.

While I have faith that we humans, as largely cooperative social animals by nature, have the capacity to identify more with our propensity for caring, sharing, and community rather than competitiveness, selfishness, and excess, the challenge is finding ways this alternative human identity can be socially encouraged. This paradigm shift is less likely under a neoliberal political system that emphasizes economic rationales and capitalistic growth and consumption initiatives that are often ecologically-unsustainable and exploitative. A just humanity is also impeded by a worldview, often cultivated by major religions, that supports a human superiority complex over all other species, as well as a legacy that privileges whiteness and masculinity.

This dominant ideology has been socially constructed over time, largely by and for those in power, to maintain status quo power structures. I share a Gramscian and Foucauldian-inspired view that this hegemonic discourse can be deconstructed to reveal it isn't natural or fixed and other identities are possible. As a main site of ideological struggle, the media are highly influential at cultivating and maintaining our political and cultural values. But the "public sphere" is largely dominated by corporate media that are subsidized by advertising and, therefore, cater to commercial interests more so than social and ethical interests. While some resistance is possible via mainstream media, activists need to circumvent corporate media and produce their own direct communication with the public, which is the focus of this book's study.

I don't view these animal rights activist campaigns primarily as social marketing (like social cause advertising for behavior change) or necessarily even persuasion (at least not in a manipulative sense), but as provocative, strategic presentations of an alternative human identity, one based more on values of humility, mutual respect, and responsibility toward our collective interest as fellow earthlings. So in discussing social change processes in this book, I draw less upon psychology, marketing/PR, or political science and more upon social movement framing theories (as part of communication scholarship). I believe social movements serve a vital role as moral innovators, encouraging society to think critically and question fundamental assumptions, spurring social institutions to start reconstructing our social practices for increased fairness and wellbeing. It's in this cooperative mindset and spirit that I offer my analysis and recommendations to fellow activists.

Acknowledgements

I am particularly grateful to the activists and animal rights organizations featured in this study – Matt Ball and Vegan Outreach; Gene Baur and Bruce Friedrich and Farm Sanctuary; Alex Hershaft and FARM; Erica Meier and Compassion Over Killing; and Ingrid Newkirk and PETA. Because this study is based on my doctoral dissertation, I owe gratitude to my committee members Tom Bivins, Carl Bybee, Michael Dreiling, and my supportive adviser and fellow critical animal studies media scholar Debra Merskin. I was so pleased that Rodopi Press offered a critical animal studies book series to help foster scholarship in this burgeoning discipline. The book series reviewers, and especially the editors Helena Pedersen and Vasile Stanescu, were highly supportive at facilitating my first book project. And I appreciate the following non-profits for offering resources to me: The Animal Welfare Trust, the Culture & Animals Foundation, and the Humane Society of the U.S.'s scholars retreat. I appreciate the support and academic freedom I receive at Georgia State University, including final editing help from graduate assistant Christopher Toula. My friends and family are also particularly supportive, especially my parents, brother and sister-in-law, and niece and nephew (Steve, Cathy, Scott, Ann, Olivia, and Grant Packwood) who enjoy animal-free veg meals with me. I was also inspired daily by my late, long-time roommate, Marley, a splotchy black and white cat, who, while not a farmed animal advocate, reminded me of cows used for dairy and veal.

Dedication

I would like to dedicate this book to the billions of individuals who are bred or stolen to be used as consumable items, and to everyone helping to create a more just and sustainable humanimality.

Chapter 1

Introduction

How much should animal rights activists foster animal rights when they attempt to convince meat-lovers to stop eating animals? In promoting veganism to a meat-eating public, animal rights activists face a classic communication dilemma that all counter-hegemonic social movements have historically faced. Should campaign messages be more pragmatic and utilitarian (ex: emphasizing reform and human self-interest) or more radical and ideological (ex: emphasizing justice, abolition, and altruism)? For vegan advocates, this means deciding between pragmatically meeting people where they are (ex: messages promoting meat reduction and farmed animal welfare) or taking them further to challenge discriminatory beliefs (ex: messages promoting animal rights and veganism). In this book I provide a pathway for the latter, what I call "ideological authenticity," where persuasive messages are grounded in the advocate's ethical philosophy to promote a transformation in worldviews not just behaviors.

While it is important and popular for activists to ask pragmatic questions such as "what communication approach works best as a means to an end?", in this book I prioritize the more fundamental question of what communication approach is most fitting and authentic for a social movement so that they construct messages representative of the transformational values they aim to instill in society. The assumption is that what is true to a counter-hegemonic movement's ideology should, in most cases, be publicly communicated as such, both to emphasize honesty and integrity in means and to achieve the desired ends of transforming discriminatory worldviews. This is inspired by Michel Foucault's (2000) endorsement of radical criticism as a necessary constructor of discursive transformation:

> For a transformation that would remain within the same mode of thought, a transformation that would only be a certain way of better adjusting the same thought to the reality of things, would only be a superficial transformation. (457)

By expressing their oppositional ideology, social movement discourse should result in people having "trouble thinking things the way they have been thought" (457).

This book confronts a similar line of inquiry in social movement framing, deciding whether appealing to an individual's self interest is counterproductive to the long-term goal of getting society to be more altruistic toward a new category of oppressed beings (Crompton and Kasser 2009). If activists seek a more altruistic society, should they emphasize altruistic values, even if that might not be the quickest path to effect some changes? For example, if an animal rights organization can convince more people to stop eating animals, or to eat fewer animals, by appealing to legitimate human health concerns, is that preferable to a moral suasion approach that appeals to people's sense of justice and empathy toward other animals? The former, self-interested health frame might be an easier or more persuasive way to get an audience member to stop/reduce eating animals, but because the frame does not fundamentally challenge humanity's hegemonic views toward other animals, the new vegetarians may see nothing wrong with supporting fur, leather, hunting, or animal experimentation. Would it not be more authentic, and even more strategic in the long run, if each separate animal rights campaign, whether it be against farms, laboratories or entertainment industries, was informed by the same, core non-speciesist ideology instead of separate appeals to human self-interest or the welfare of that species in that instance?

This book arrives at a key moment in the animal rights movement where many agree that billions of intensively farmed animals deserve the movement's primary attention (Ball and Friedrich 2009; Torres 2008), but internal debates over strategy create a mixed external message about precisely why and how the public is to help end this exploitation or whether, perhaps, just to mitigate it. In this book I propose my thesis that the ideal messages are ones that both culturally resonate with people and openly ask for the kind of radical change in speciesist worldview that is necessary to promote all animal rights issues in the long term. This would mean constructing vegan campaign messages that not only convince people to avoid consuming animal products but that do so in a way that enables people to respect other animals as fellow sentient beings with the right to live free of exploitation (more pro-fairness rather than just anti-cruelty). In this way, campaign messages connect with broader and more fundamental issues of ethics, justice, rights, and ecology in promoting a deconstruction of the human/animal and culture/nature dualisms so that humans embrace their own animality and consider establishing a humbler, fairer, and more sustainable place in the world.

Themes & Theses in This Book

To examine the place for ideology in the framing process of a counter-hegemonic social movement, I look at the specific case of how five animal rights organizations currently frame issues and values in their advocacy for farmed animals. I elected to study the five animal rights organizations that I deemed most actively engaged in food advocacy in the United States: People for the Ethical Treatment of Animals, Farm Sanctuary, Compassion over Killing, Farm Animal Rights Movement, and Vegan Outreach. Through two rounds of interviews with the organizations' leaders and a rigorous analysis of their "go veg" campaign materials (websites, brochures, videos, advertisements, and collateral materials), I examine how these organizations construct farmed animals and their issues, paying particular attention to the construction of the *human* animal and our values regarding other animals.

The reason for this focus on human values is that, even though one can assume that animal rights campaigns will show nonhuman animals as sentient beings who suffer greatly, the real question becomes whether human society cares enough about other animals to cease exploiting them and to question humanity's sense of entitlement. We humans are the only species who can improve the animals' situation, as we endorse their exploitation legally, financially, and socially through the common practice of farming, fishing, and eating animals. Therefore, I examine the positioning of the human subject in terms of what values activists are suggesting that humans do or should possess.

By analyzing vegan campaign materials and interviewing organization leaders, I answer questions such as:

- How is the American animal rights movement defining the core problems and solutions regarding "food animal" issues?
- To which values are activists appealing and why?
- In what ways and to what extent are they promoting altruistic values as opposed to appealing to human self-interest?
- How can appeals to human health and environmentalism also reflect an animal rights worldview?
- To what extent is the animal rights movement challenging speciesist discrimination (injustice) and the human/animal divide?
- How do animal rights leaders explain and justify their framing choices in terms of ethics and ideology?

All of this description leads to a prescriptive component where I closely evaluate the implications of framing choices for animal ethics, communication strategy, and the animal rights movement. I discuss my prescriptive assessment at the end of most chapters, especially in the final chapter, which includes my detailed recommendations for frames that are potentially the most transformative and supportive of animal rights ideology, such as emphasizing justice, respect, freedom, life, and a shared animality more so than the current emphasis on factory farm cruelty. This authentically aligns animal rights theory with advocacy practice in a way that also strategically incorporates both environmental ethics and human rights and merges nature and culture. I believe that a societal progression toward animal liberation requires a transformation in humans' conception of *themselves* in relation to other animals, as a more integrated part of the animal kingdom.

The Unique Contributions of This Book

Most books on vegetarianism try to convince the readers to adopt the diet or give them nutritional and culinary tools to navigate this alternative lifestyle, but my book is about *how* we can create a needed global shift to a plant-based diet, speaking directly to activists and those sympathetic to the animal cause. I provide advocates with the rhetorical tools they need to make this happen. Related books on the market tend to focus on various aspects of vegetarianism, or animal ethics, or activism, or strategic communication, but I wrote this book to unify *all* of these pertinent topics. Therefore, it offers a unique, useful, and timely contribution to knowledge in social movement studies, communication and framing studies, public relations, environmental communication, food studies, advocacy ethics, animal ethics, and critical animal studies.

As part of a critical animal studies (CAS) book series, my interdisciplinary and applied approach fits well with many of the tenets of CAS (Best et al. 2007). One of the pioneers of CAS, Steven Best (2009), described the field's goal as seeking to "breakdown and mediate oppositions between theory and practice, college and community, and scholarship and citizenship, in order to make philosophy (in a broad sense) again a force of change and to repatriate intellectuals to the public realm" (12). To facilitate this engaged scholarship, I am using animal ethics and communication studies as a platform for advising animal rights activists in constructing their critical discourse on food so that scholarship advances tangible social change. Often books on advocacy are designed either for communication practitioners (offering anecdotal examples and useful how-to guidelines) or primarily for scholarly audiences (offering theoretically-rigorous critique but little guidance). As both an activist and scholar, I designed this book to be philosophically deep yet practical, offering concrete, nuanced guidance on communication strategy while remaining theoretically grounded. It is descriptive in terms of critically analyzing what *is* being communicated in the contemporary "go veg" campaigns of major animal rights organizations, but it is also theoretical and prescriptive in recommending what *should* be communicated for long-term social transformation. Additionally, my anti-speciesist, pro-vegan ideological stance fits with critical animal studies beliefs in openly advocating for human and nonhuman liberation rather than just studying animals as a theoretical curiosity. Describing CAS, Best (2009) said "It openly avows its explicit ethical and practical commitment to the freedom of well-being of all animals and to a flourishing planet. It opposes all forms of discrimination, hierarchy, and oppression as a complex of problems to be extirpated from the root, not sliced off at the branch" (12).

CAS also promotes direct action and grassroots activism. While I'm studying well-organized national nonprofit groups with funding budgets in the hundreds of thousands (or even millions) of dollars, my strategic advice is also applicable to grassroots and direct action groups with meager funding in the hundreds of dollars.[1] CAS can be described as radical, and my strategic rhetorical advice promotes a radically transformative worldview, but I don't necessarily dictate the level of radicalism or conservatism of the tactics themselves in terms of delivery, so long as they are representative of the values the animal protection movement seeks to instill in society. My study is not really about abolition versus reform *tactics,* as it is more about identifying a practical abolitionist *rhetoric*. It assumes an abolitionist stance is preferred and seeks to better understand how this can be rhetorically framed in ways that are ideologically authentic but still culturally resonant to a public more aligned with welfare than rights.

CAS scholarship promotes a political economic critique of capitalism and connects speciesism with other discrimination systems such as racism and sexism (Best et al. 2007). While these elements are more tangential to my analysis, I do think my thesis helps to align all movements against instrumentalism when I advocate that activist rhetoric should prominently debate enslavement, exploitation, and injustice to create a more altruistic, less self-interested society in which everyone matters. While my book does focus on the case of the *animal rights* movement's communication challenges, the findings can be abstracted to apply to the common dilemma faced by almost *any* counter-hegemonic movement in determining how they can be critical of the status quo while still remaining resonant and effective at creating major social change, both behaviorally and ideologically. And although the book's emphasis is on *American* culture (or Western culture more broadly), I hope that activists across the globe can adapt and apply its basic advocacy lessons in their own cultures. While cultural context varies, I believe many values and ethical concepts are universal.

Social Significance of Veganism and Animal Rights as a Topic of Inquiry

While veganism can seem like a narrow topic just related to a dietary preference, in actuality it has broad-based implications for everyone on the planet. People are starting to recognize its impact, as food and

---

[1] When I ran the all-volunteer Vegetarian Society of Southwest Florida in the 1990's,

agricultural reform continues to rise in popularity for reasons of ethics and welfare, public health, and sustainability, which is particularly relevant in this era of global warming, mass extinction of species, and industrialized farming and fishing. The scope of the problem is massive.

Largely anthropocentric, communication research has not given adequate attention to the topic of media and advocacy related to animal farming and a meat-based diet. Perhaps this is due to the fact that mainstream media discourse rarely includes an animal rights perspective challenging the basic premise that it is acceptable to raise and kill someone else for food. The media usually construct food around the concepts of pleasure, nutrition, or economics, not around the concepts of justice, ethics, and sustainability (Freeman 2009a). This humanist media bias means that the animal advocacy texts I study here serve a much needed function of openly critiquing the mainstream food discourse. While admittedly a minority voice, the animal rights advocates' attempts to make production and consumption of animal products an ethical issue represent an important challenge to social norms and basic ideals about who it is morally acceptable to kill and who pays the cost for America's food choices.

I hope the construction of knowledge in this book raises the status of veganism as a legitimate academic topic and helps remedy the lack of attention nonhuman animals receive in academia, particularly in communication studies.[2] This is especially important given the profound real-world effects that animal advocacy could potentially have on: the billions of nonhuman animals killed annually in the food industry; cultural acceptance for animal rights and animal welfare; promotion of anti-instrumental and altruistic moral values in society; equitable food distribution and human health; and environmental protection and sustainability.

Fortunately, farmed animals have been made a priority by the animal protection movement over the last decade. This is due to the acknowledgment that animal agriculture and commercial fishing are responsible for the overwhelming majority of nonhuman animals killed by humans in the United States (Ball and Friedrich 2009). The relevance of the food issue to animal advocacy is increased by the fact

---

[2] The burgeoning sub-discipline of environmental communication has begun to remedy the humanist bias in communication studies and allows for the respectful consideration of the interests of nonhuman life.

that meat-eating involves the majority of the public more directly than other animal issues because, while most Americans do *not* wear fur, participate in sport hunting, or conduct animal experimentation, approximately 95% of Americans *do* eat flesh taken from animals (and about 97% eat milk and eggs taken from animals) (Stahler 2011). To meet this consumer demand, the animal agriculture industry kills over 9 billion land animals annually in the United States alone (Humane Research Council 2011).[3] Additionally, an estimated 17 billion animals from the sea are sold for American food, not including the approximately 25 percent additional lives lost and wasted as "bycatch" (Singer and Mason 2006, 112). If one includes sea animals in addition to land animals, American omnivores are responsible for the killing of more than five million nonhuman animals every hour of every day.

Humans' food choices are a key issue for *all* nonhuman animal species because if people continue to breed, grow or capture, and kill other animals for food when it is unnecessary for survival, then the animal rights movement will not be able to gain significant rights for animals in any other area in which animals are commonly exploited (Francione 1996; L. Hall 2006a). For example, it makes sense to me that *omnivorous* humans would endorse animal experimentation to potentially save human lives, use animal fur/skins for warmth or fashion, and sacrifice nonhuman animal habitats for anything from drilling for oil/gas to developing new golf course communities. When society allows the needless killing of nonhuman animals for food every day, this routine meat-eating ultimately makes nonhuman life cheap in comparison to human life, putting all animals at risk of human exploitation for selfish ends.

What I deem as America's largely selfish practice of animal consumption also has negative repercussions for *humans* globally, as animal-based foods are related to issues both of nutritional excess (obesity and disease) and deficiency (malnutrition and starvation). While many in America are suffering from lifestyle-based diseases (due in part to diets high in animal-based cholesterol and saturated fat), millions of people worldwide die of hunger-related causes annually, due in part to inequitable food distribution (Pollan 2006; Singer and Mason 2006). America produces enough plant crops to feed starving

---

[3] The death toll is even higher when one includes the millions of male chicks killed at egg hatcheries and the millions of animals who die on the farm and in transport.

humans worldwide, but the nation inefficiently uses most of its crops, particularly grain and soy, to fatten farmed animals, which also unsustainably uses other life-sustaining natural resources (Humane Society International 2011; Robbins and Patton 1992). Confined animal feeding operations, also called "factory farms," and all the crops required to feed these billions of animals, cause pollution and use significantly higher amounts of resources such as soil, water, land, and energy than does a plant-based diet (Singer and Mason 2006).

The unsustainability of animal agribusiness causes environmental problems for all life on earth. Magazine editors at the Worldwatch Institute (2004) concluded:

> The human appetite for animal flesh is a driving force behind virtually every major category of environmental damage now threatening the human future – deforestation, erosion, fresh water scarcity, air and water pollution, climate change, biodiversity loss, social injustice, the destabilization of communities and the spread of disease. (12)

This indictment is seconded by a United Nations report from their Food & Agriculture Organization (FAO 2006) who described animal agriculture as "one of the most significant contributors to today's most serious environmental problems" (para. 2), including global warming, estimating that raising animals for food generates 18 percent of the world's greenhouse gas emissions, proving even more damaging than transportation.

The human demand for food from the sea has also caused a serious crisis in ocean life. The Pew Charitable Trust's Environmental Group (2007) warned that "90 percent of the world's large fish have disappeared" (7), particularly due to overfishing and trawler nets that "indiscriminately strip life from the sea" (6). And irresponsible aquaculture practices (fish farming) cause pollution, spread disease and exotic organisms into wild populations, and inefficiently raise carnivorous fish, such as salmon, who eat more protein than they produce.

Environmentalists often advocate for less inefficient animal agribusiness practices where deemed sustainable in certain bioregions, such as raising grass-fed cows on land not usable for plant agriculture (Pollan 2006). While environmental organizations tend to advocate for local and organic foods more so than plant-based (Freeman 2010c), author James McWilliams (2009) argued that the environmental benefits of local and organic foods are exaggerated (or oversimplified), but does see the environmental rationale for greatly decreasing consumption of animals. Vasile Stanescu (2010) considered a narrow focus on food miles and localism "dangerous" (29) if it overrides other environmental and justice-oriented concerns in raising fellow animals for food, and he advocates for a truly green food movement that blends environmentalism and veganism. As the European Union (and the United States to some extent) begins to tighten its domestic environmental and animal welfare regulations, a continued demand for animal-based foods sends factory farms to developing countries, exporting the environmental, health, and welfare problems across the globe (Nierenberg 2003). The topic of this book has never been more important, since global consumption of animal products, if left unchecked, is expected to double halfway through the century, especially as developing nations are targeted as an emerging market for animal foods (FAO 2006; Twine 2010).

The Structure and Content of This Book

PART I (Chapters 2 & 3): Because the articulation of animal rights ideology and values are so central to this book, I set the stage for this discussion by explaining fundamental aspects of animal ethics and communication theory in Chapters Two and Three. Chapter Two draws upon philosophers to provide context on Western society's outlook on themselves in relation to nonhuman animals, including ethical views specific to vegetarianism and current consumer attitudes about meat consumption and animal farming. This chapter includes my theory that the deconstruction of the human/animal dualism is essential to animal rights ideology. And Chapter Three incorporates communication scholarship and the lessons of rights activists to outline strategic communication challenges and guidelines specific to all social movements, drawing upon rhetorical debates over ideological purity versus expediency in the early movements for women's rights and abolition of slavery. I apply this to recent animal rights movement debates over authenticity versus pragmatism in deciding how best to frame abolition, welfare reform, and vegetarianism.

PART II (Chapters 4, 5, 6 & 7): The second part of the book comprises description of my empirical analysis, explaining how the five animal rights organizations that I studied framed food issues and why. While the bulk of each chapter is filled with rich description (direct quotes and examples), at the end of most chapters I share my own evaluation of the implications of the framing choices for the animal rights movement, considering these choices in context to the recommendations of communication and social movement scholars.

Chapter Four introduces the research study method and profiles the five organizations before detailing the findings of the first research question, where I categorize how animal rights organizations diagnosed problems with the production and consumption of animals and proposed solutions to these problems. Findings explain how activists framed problems around farmed animal cruelty and commodification, human and environmental harm, and unnecessary killing. Solution frames suggested consumers eat a total (or largely) plant-based diet, and some proposed industry welfare reforms. I suggest ways that proposed problems and solutions could more logically align around justice through veganism.

Chapter Five identifies sixteen values that the animal rights organizations appealed to when framing the problems and solutions, as part of their attempt to motivate people to identify with the cause and to see veganism as an ideal fit for their values – values such as: compassion, sentience, moral consistency, desire to make a difference, choice, pleasurable and convenient food, belonging, life, concern for fellow human beings, honesty, American populism, naturalness, freedom, and American pride. Chapter Six continues the discussion on values, categorizing the activists' frames as either primarily appealing to human self-interest or altruistic values (toward nonhuman animals or anthropocentric altruism), while also noting the way altruism and self-interest overlap in frames emphasizing the mental health benefits of living your values as a vegan. Overall, I found most campaigns did focus on nonhuman issues more so than human-centered issues. I discuss how, to further reflect this dedication to nonhuman animal protection, activists could place greater emphasis in environmental frames on the negative effects of animal agriculture on wild/free animals and their habitats (rather than primarily on human health).

In Chapter Seven we move away from my empirical study of the activist materials to my 2008 interviews with the five animal rights organization leaders: Matt Ball (Vegan Outreach), Gene Baur (Farm Sanctuary), Bruce Friedrich (PETA), Alex Hershaft (FARM), and Erica Meier (Compassion Over Killing). Here these prominent activists get a chance to explain their strategic communication choices and share their expertise, giving rationales for why they framed issues as they did. To structure their ethical decision-making approaches as either deontological/authentic or more pragmatic/utilitarian, I draw upon classic social movement framing debates from Chapter Three. Strategically, leaders applied both deontology and utilitarianism decision-making in choosing to prioritize altruism to nonhuman animals more often than human self-interest, but most leaders favored utilitarianism in choosing to privilege animal *welfare* over animal rights for wider appeal.

PART III (Chapters 8 & 9): While part two is largely *de*scriptive, part three is primarily *pre*scriptive. In Chapter Eight I share my latest round of interviews in 2012 with animal rights organization leaders (Matt Ball, Gene Baur, Bruce Friedrich, Alex Hershaft, Erica Meier, and Ingrid Newkirk) to compare and contrast their interesting insights on their latest campaigns, the future of animal activism, and what strategic communication approaches this necessitates.

And in the final chapter, Chapter Nine, I provide my own recommendations for how the animal rights movement should frame problems and solutions using values-based appeals. I conclude that while some activist messages in my study did support animal rights (promoting veganism and respect for nonhuman animals' subject status), many frames used animal *welfare* ideology (emphasizing suffering on factory farms) to achieve animal *rights* solutions (veganism), conservatively avoiding a direct challenge to the dominant human/animal dualism. Therefore I provide examples and nuanced recommendations for framing animal foods as a problem based on injustice, first and foremost, while still problematizing suffering and environmental destruction. I suggest engaging the audience as both consumers and citizens to explain their culpability and their capability toward individual and collective solutions, including 1) appreciating the mutual subject status of all animals (including humans); 2) eating a plant-based diet; and 3) working collectively to create a less speciesist society. Infused through all of this should be strategic appeals to values such as: fairness, respect, life, freedom, integrity, honesty, naturalness, responsibility, moderation, community, diversity, caring, compassion, peace, sharing, humility, accountability, making a difference, and health. For added depth, I consider my recommendations in context to their fit with communication theory, acknowledging limitations and highlighting strengths.

WORD CHOICE

In this book, you'll notice I use the term *farmed* animal instead of *farm* animal. I do so to foreground the fact that farming is something we do to these individuals – something we force upon them. This is inspired by other scholars who have respectfully chosen to use the term *enslaved person* instead of labeling someone a *slave* (Brown 2004; Dunayer 2001; Howard 2006). Additionally, to help linguistically deconstruct the human/animal dualism and emphasize that humans are animals, I often use the term *nonhuman animals* when it is necessary to distinguish all members of the animal kingdom besides humans. Therefore, when I just say *animal(s),* it typically includes humans too. When I'm speaking only of humans, I sometimes refer to them as *human animals* or with the combination word *humanimal.* This is part of my rhetorical attempt to embrace humanity's innate animality and attempt to overcome its connotation as an insult.

Part I:
Overview of Animal Rights, Vegetarianism, and Communication

Chapter 2

Ethical Views on Animals as Fellows and as Food

It's useful to clarify animal rights ideology (as distinct from animal welfare) as it serves as a philosophical basis for this book's future analysis assessing the ways in which the messages from animal rights organizations are informed by or supportive of this ideology. I believe you really cannot design effective advocacy messages until you truly understand and interrogate your own ideology (including assessing its limitations as well as strengths), as that deeper understanding of your beliefs and motivations needs to shape your surface-level rhetorical choices.

So in this chapter I cover broad territory in reviewing Western society's current beliefs about fellow animal beings, including views on nature, ethics, and using other animals for food. I begin with a brief overview of the history of animal rights activism in the United States, including activism on behalf of farmed animals, which helps situate this book's focus on animal rights as a movement. Then I summarize modern animal rights philosophy and the rhetorical challenges of deconstructing the false human/animal dualism at the heart of speciesism. Narrowing my focus to food issues, I end the chapter with an intriguing review of vegetarian ethics and the status of vegetarianism and meat consumption today. This includes discussing the relationship between meat and other forms of oppression like patriarchy, discussions on what diets reduce the most suffering, how to reconcile vegetarianism with predation/hunting in nature, and society's psychological need for deception in communicating about meat.

DEVELOPMENT OF ANIMAL ACTIVISM IN UNITED STATES

The United Kingdom is thought to have pioneered the Western humane movement in the early 19th century with Richard Martin's founding of the Society for the Prevention of Cruelty to Animals, which was formed to enforce new anti-cruelty laws protecting farmed animals (Beers 2006). In 1866 the American Society for the Prevention of Cruelty to Animals (ASPCA) was founded by Henry Bergh. His first conviction was a butcher who had mistreated farmed animals (Beers 2006). Early ASPCA campaigns sought to improve the conditions for working carriage horses and farmed animals in the slaughterhouse as well as to stop hunting and animal experimentation. While some states did pass anti-cruelty laws earlier in the 19th century, the nation did not have a federal anti-cruelty statute until 1871 when it passed the "twenty-eight hour law" to improve the welfare of farmed animals transported by rail to slaughter. But protecting farmed animals drew less public support than protecting companion animals, so pet shelter and rescue work started to dominate the humane movement in the late 19th and early 20th centuries. Post World War One (WWI) did see other campaigns such as those against fur and circuses (Beers 2006). The post World War Two (WWII) years saw a growth in professional welfare organizations like the Humane Society of the United States (HSUS), Friends of Animals, and the Animal Welfare Institute, who focused on changing institutions not just individuals (Jasper and Nelkin 1992).

There were divisions within the animal protection movement, however. The American Humane Association (AHA), which formed in 1877 to unify the movement, ironically caused a rift between radical and conservative activists in the late 19th century, as it took a conservative welfare stance of working with industries like ranchers (Beers 2006). More militant activists left and formed their own rights groups, such as the American Anti-Vivisection Association started by Caroline Earle White. By the post WWII era, even other welfare groups, such as the HSUS campaigned for more stringent humane reforms in slaughterhouses, as the HSUS considered the AHA's slaughterhouse monitoring to be too weak. Welfare organizations sometimes used rights language but ultimately they took an instrumentalist view that weighed human interests higher than the interests of other animals and did not promote species equality. Welfarists and rightists argued over language in reform bills, but the welfarists won and passed the Humane Slaughter Act in 1958 and the Laboratory Animal Welfare Act in 1966 and 1970. The animal protection movement was largely dominated by welfare groups until the mid 1970's (Beers 2006).

In its origins, the animal protection movement borrowed activist strategies from the abolitionist and women's rights movements, such as using moral suasion to expose the reality of injustices (Beers 2006). Since animal activists believed people were not willing to give up their superior status over other animals, the early movement leaders often used an anthropocentric approach and attached humane reform to human self-interest. For example, the Humane Slaughter Act was also touted as a public health reform. Many organizations, especially the ASPCA under Bergh's leadership, attracted media attention by staging protests and using shocking visual images of cruelty. Historian Diane Beers (2006) claimed that newspapers covering the emerging movement in the late 19th century often ridiculed activists, particularly Bergh, as sentimentalists. But the news of that century eventually did show some moral outrage and sometimes compared human and nonhuman animal slavery.

The modern day animal protection movement was inspired by Peter Singer's 1975 book *Animal Liberation* (Jasper and Nelkin 1992). In the 1970's, activist Henry Spira individually led animal rights campaigns, but by the 1980's, national animal rights groups formed, such as People for the Ethical Treatment of Animals, In Defense of Animals, the Animal Liberation Front, and the Animal Legal Defense Fund, all of which are still active today. By the end of the 1980's, there were several hundred animal rights groups and several thousand welfare groups, mainly local humane societies. Membership in national groups rose drastically in the 1980's – mainly from educated, city-dwelling, non-religious women who had companion animals[1] (Jasper and Nelkin 1992). To give a perspective on the rapid development of the humane movement, at the turn of the 20th century there were about 700 animal protection organizations. One century later, the number increased to approximately 7,000 organizations with over 10 million members (Beers 2006, 3).

[1] This just described my own demographic.

The animal protection movement of the late 20th century drew from the ideologies of the feminist and environmental movements to critique instrumentalism, the institutionally-sanctioned exploitation of others as a means to an end (Jasper and Nelkin 1992). The anti-instrumentalism of the New Left questioned capitalism's growth imperative and its emphasis on the material instead of the moral. Rights rhetoric burgeoned in many movements in the 1970's, including animal rights. Sociologists James Jasper and Dorothy Nelkin (1992) claimed the animal rights fundamentalists were more successful than welfarists at attracting members and formulating issues because they used strong visuals and moral language that was more dramatic and energizing. However, the authors critiqued the fundamentalists' message as too polarizing due to a demonization of opponents as enemies.

The moral language of rights had radicalized the animal protection movement in a matter of a few decades. Jasper and Nelkin (1992) explained:

> Their leaders have linked philosophical arguments about the exploitation of animals to prevailing social concerns: the mistrust of science and medicine, the disaffection with big business and commodity culture, the disillusionment with bureaucracy and expertise, and the resistance to domination so important in feminist critiques. (170)

As philosophy and ethics are so inherent to animal activism and the modern message strategies examined in this book, in the next section I take a step back to examine Western society's views and attitudes about themselves in relation to other animals.

WESTERN THOUGHT ON OTHER ANIMALS

Since its birth in ancient Greece, Western philosophy has largely focused on a privileging of the human subject. David Schmidtz (2002) claimed that philosophy has historically been an examination of the following three anthropocentric projects: determining human's essence, specifying how humans are different from all other species, and specifying what makes humans morally important. Most philosophies are not only focused on humans, they also assume humans are morally superior. Paul Taylor (1993) claimed that the following three traditions were mainly responsible for constructing the idea of human superiority: Greek humanism and its privileging of man's rationality; the

Cartesian dualism which divided animals into humans who possess a mind and a soul and other animals who only possess a body; and the Judeo-Christian "great chain of being" that ranks god first followed in descending order by angels, humans, animals, plants, and inanimate objects. Peter Singer chastised his own field of philosophy for its inherent anthropocentrism, claiming it had failed both to challenge accepted beliefs and to justify its assumption of human dignity (Linzey and Clarke 2004). Singer (1990) highlighted how convenient this uncontested anthropocentrism is by asking, "Why should we not attribute 'intrinsic dignity' or 'intrinsic worth' to ourselves? Fellow humans are unlikely to reject the accolades we so generously bestow on them, and those to whom we deny the honor, are unable to object" (239).

To contest anthropocentrism, the scholarly field of human-animal studies began 40 years ago in analytic philosophy, particularly developed by Peter Singer and Tom Regan in the 1970's, and has since branched out to be interdisciplinary (McCance 2013) in what Steven Best (2009) and others called the "animal turn" (9) in academia. Critical animal studies exemplifies an explicit commitment to fighting discrimination and oppression, adding species to the intersectional analysis of race, gender, and class and emphasizing the role of capitalism in exploitation of all these entities (Best 2009; Twine 2010). Richard Twine (2010) defined animal ethics scholarship as "shallow" (21) if it is limited to a philosophical approach rather than embracing the interdisciplinarity of animal studies and if it fails to question core humanist assumptions about liberal individualism and anthropocentric cognitive comparisons.

What follows is animal studies scholarship, from philosophy and other disciplines, that focuses on justice issues in our relationship with nonhuman animals.

Modern Animal Rights Views

In the 1970's, Peter Singer (1990) built on the concerns of past utilitarian philosophers, Bentham and Mill, to propose that all sentient animals should have their like interests given equal consideration. He defined sentience as the ability to suffer and experience happiness, both of which are key concerns in a utilitarian calculation of maximizing pleasure versus pain. Sentience, even more so than intelligence, was the most morally relevant trait a being possesses, as sentience is the common denominator humans respect most in each other. To prove this, Singer used marginalized case examples showing that humans still care about the interests of sentient, developmentally-challenged humans, regardless of their intelligence level.

If people argued that the morally relevant trait was simply being human, instead of sentience or intelligence, Singer (1990) accused them of species discrimination because they failed to provide a reason for the moral relevancy of species in ignoring the like interests of others. While Singer admitted that it initially seems logical to claim that favoring the interests of one's own species is similar to how one naturally favors the interests of one's own family group, he revealed this rationale's inconsistency by saying it would lead to racism or sexism if applied to showing favoritism for one's own racial or gender group. Singer argued that when humans elevate the status of their own species, they effectively lower the status of others, making humans guilty of species discrimination. To label this unjust bias against nonhuman animals, he used the term "speciesism" (Singer 1990, 6), which he credits to psychologist Richard Ryder.

Speciesism is linked with racism and sexism, as there are strong parallels in how women and people of color have been discriminated against by being compared to lowly and irrational animals (Adams 1990; Singer 1990; Spiegel 1996). While Mary Midgley (1984) conceived of race as a more arbitrary category, biologically-speaking, than species or gender, she agreed that rights movements on behalf of race, gender, and species are ultimately all working toward the same goal of defeating "unfairness" or "unreasonable biases" (101). Biases enable hierarchies, which often lead to mistreatment, where the "superior" group feels justified sacrificing the major interests of the "inferior" group to satisfy their own minor interests (Singer 1990).

In addition to Singer, Tom Regan was one of the first contemporary philosophers to ask for moral consistency in humans' dealings with other animals, paying nonhumans similar courtesies as are shown to humans under a human rights model (Linzey and Clarke 2004; Regan 1983). Regan (2004) emphasized rights over suffering by declaring humans should respect the right to life and liberty of all individuals who are subjects of a life, regardless of species. He explained that what is shared by many animals, particularly mammals, is that "we are each of us the experiencing subject of a life, a conscious creature having an individual welfare that has importance to us whatever our usefulness to others," (Regan 2004, 185) and so we all have inherent value. Under Regan's duty-based viewpoint, it is immoral to treat those with inherent value as though they are just a utility; all who have inherent value have it equally. Therefore, the fundamental wrong in society is humans' systemic view of other animals as resources, and Regan called for abolition of humans' industrial exploitation of them.

The difference between Regan's view and Singer's is that the former is more of a rightist while the latter is more of a welfarist, according to environmental philosopher Gary Varner (1998). But I discern overlapping elements between both philosophies, as they each seek fairness in extending the egalitarian notion of respect that society has for all humans out to other fellow sentient, conscious beings. They both differ from a more broad-based philosophy like Albert Schweitzer's reverence for all life, as they exclude plants and less conscious animals, such as oysters. And with both Regan and Singer, *levels* of sentience and individual consciousness still come into play, as species thought to more clearly possess these traits (which we assume to be human-like) become more deserving of moral relevance.

Mary Midgley argued that our concern for nonhuman animals should be based on humans exercising compassion, not based on the other animal's interests or rights, as compassion is less abstract and does not ask that all animals be treated equally (Jasper and Nelkin 1992). Compassion does not require the anthropomorphic identification with other animals based on similar mental states of consciousness. One simply needs to feel sympathy to avoid causing others to suffer. Building off this notion of sympathy, some feminists, such as those using Carol Gilligan's ethic of care, find Singer and Regan's arguments too individualistic, abstract, and rationalistic. They prefer to emphasize kinship, community, relationships, and connection, believing we should act not out of duty but out of sympathy, love, and caring for fellow animals (Donovan and Adams 2007).

Kinship and community are also key components to philosopher Gary Steiner's (2008) moral philosophy. He asked us to be empathetic and "acknowledge their plight and their prospects in a world that has been dominated by human beings" so that we can start to "identify with animals, to see ourselves in them and them in ourselves" (137). He regretted that humans have become too distant from animals and nature, subjugating our shared experience with animals in pursuit of aligning ourselves with the divine. Instead, he encouraged humans to "acknowledge our mortality and the fundamental limitation that we share with animals" (137), and to cultivate a "sense of dwelling alongside animals in a cosmic whole" (125). Steiner advocated a *cosmic justice*, which he defined as "to let animal beings be in such a way that we no longer project upon them a diminished reflection of our own image but instead value their mortality as we value our own" (163).

Paola Cavalieri (2006) declared that the Cartesian era of viewing animals as automata is over and Western society is now back to the debate that is over 2,000 years old, the "original Greek appraisal of the worth of other animals" (66). Society is moving beyond the limited arguments over cruelty and pleas for more compassion and is challenging the idea that nonhumans should be enslaved. For the first time in history, using philosophical means, "it is now possible to defend the idea that animal lives have value" (66).

The following section deepens the discussion of modern philosophies relating to human's relationship with nonhuman animals. It centers upon the rhetorical challenges activists face in moving beyond the anthropocentric worldview that separates humanity from "the animal."

Deconstruction of the Human/Animal Dualism[2]

The root of Western philosophy's justification for its discrimination against nonhuman animals is primarily based upon the socially-constructed human/animal dichotomy. Deconstructing this harmful dualism starts with using more respectful, inclusive language that avoids misnomers like "humans and animals" and reminds humans of their own animality (Dunayer 2001). This can be accomplished, albeit imperfectly, by using terms such as *nonhuman animals* and *other-than-human animals* to describe all animals who are not human, and using *human animal* or *humanimal*[3] to describe only humans. When wanting to describe all animals including humans, one can say *all animals, animals (human and nonhuman),* and *sentient beings*. A major rhetorical challenge in acknowledging our humanimality is the negative connotation around the term *animal* as an insult when applied

[2] This section is a summary of a published book chapter of mine in *Arguments About Animal Ethics* (Freeman 2010a).

[3] *Humanimal* is a term I first saw used by William J. T. Mitchell in the forward to Cary Wolfe's (2003) book. While the term is admittedly cumbersome, I like that it blends *human* with *animal.*

to a human. Indeed, there is a long history of those in power using animal labels to belittle human groups for purposes of hierarchizing, marginalizing, and oppressing (Adams 1990; Spiegel 1996). Therefore, the term *animal* will need to become associated with more dignity than it traditionally has, which is a goal of the animal rights movement.

Another weakness to highlight in the false human/animal dichotomy is an inability to neatly define the human border that separates humans from all other animals, at least not in a morally relevant way (Lawrence 1995; Ucko 1988). Moral relevancy is generally determined according to *mental* criteria/capabilities, yet species categories are primarily based upon *physical*/biological traits (Elstein 2003). Singer (1990) claimed that humans value sentience as the mental criteria that distinguishes moral relevancy, and Regan (2003) proposed the key criteria was being a conscious subject of a life. Many animal species possess the mental criteria of sentience and consciousness that warrant their being considered morally-valuable individuals, despite any physical differences based on species membership (Bekoff and Pierce 2009).

Yet, even if humans start to believe that many nonhuman animals have moral value, our call to protect them may, ironically, reinforce human superiority, thus separating us further. By this, I mean that a major tension exists in animal rights when advocates base requests upon the notion that humans should be so compassionate and "humane." The noble concept of "humanity" or "humane-ness" may privilege humans with a certain ethical status presumably not found in other animals or nature (hence the word *human* being embedded in the word *humane,* implying that we are the only moral being) (Dunayer 2001; Freeman 2010a). I argue that humans' ethical system is highly complex when compared to that of other animals, but this high level of sophistication appears to be an evolutionary necessity to restrain humans' special propensity for excessive harm. I believe that the one relevant trait that does seem to distinguish the human species among most other animal species is our ability/propensity to do most things (both productive and destructive) to *excess* of what is natural or needed[4] (Freeman 2010a).

[4] I am essentializing the human species here, but I recognize that cultural and economic factors affect the ways and extent to which most human societies are excessive.

In comparison to human culture, nature is not necessarily immoral, as nature has its own principles that promote moderation and cooperation (Callicott 1993; Leopold 2003; Rolston 1993). This is especially true among social animals, as many social animals have their own guidelines that determine morality for their species' culture and often choose to exhibit ethical traits such as fairness, empathy, trust, and altruism (Bekoff and Pierce 2009). Therefore, when animal rights organizations promote animal rights on ethical grounds, they should take care not to insinuate that ethical principles are limited to the realm of humanity or are "humane," as that might unintentionally reinforce the problematic human/animal dualism and related notions of human superiority that lead to discrimination against nonhumans in the first place.

Besides the notion of being "humane," humanism is arguably also apparent to some degree in Singer's and Regan's theories that use a *shared* trait between human and nonhuman animals, such as sentience and consciousness, as a reason to include nonhumans in our sphere of moral concern. However, before chastising animal rights for implicitly promoting humanism, one must determine whether the activist's line of argumentation is based more on the desire to *build nonhumans up* in the "noble" likeness of humanity or based more on the desire to *knock humans down* off their self-constructed moral pedestal, encouraging them to embrace, instead of repress, their innate animality. The latter approach of asking humans to embrace their humanimality is perhaps a more philosophically rigorous approach to promoting animal rights, but it is not as pragmatic as the more humanist approach of proving nonhuman animal *likeness* to humans.

A philosophical problem with the tactic of emphasizing that nonhumans share many valued "human" traits is that it runs the risk of reducing other animals to lesser categories of "sub-humans." While other species are different, they are by no means failed or lesser versions of humans (Balcombe 2006; Bekoff and Pierce 2009; S. R. L. Clark 1994). Similar to civil rights messages promoting diversity and multiculturalism and environmental movement messages promoting biodiversity, I believe animal activists should ask people to respect *diversity* between animal groups and individuals. But many animal activists may not want to nor feel it is truthful to fully concede that humans and other animals are so completely different. Therefore, I contend that the best philosophical position upon which to base an animal rights message may be a blended one that embraces both the fundamental commonalities that provide kinship and the specific differences that provide diversity. I suggest that the base, connecting trait between us be a sense of consciousness – our shared status of being "subjects of a life" (Regan 2003); it is broad enough to include principles of both sameness and difference.

The next section narrows the discussion of human-animal relations to issues of food and farming.

WESTERN VEGETARIAN ETHICS

In opening their book *Ethical Vegetarianism: From Pythagoras to Peter Singer*, editors Kerry Walters and Lisa Portmess (1999) situated the specific ethical dilemma of eating other animals within the broader ethical debate of nature versus culture. The editors questioned: whether humanity's purpose was to amend nature or to follow it, how humans should view predator and prey relationships and determine which one they are, and how humans should reconcile the fact that some suffering and death is necessary to bring about a renewal of life.

Throughout history, all pro-vegetarian writers have shown a concern for the suffering that humans cause other animals, specifically suffering that is deemed unnecessary (Walters and Portmess 1999). Writers often asked humans to stop repressing their pity and to bear witness to the horrors of the slaughterhouse. While writers from previous eras focused more on questions of moral purity and how to lead a good life, contemporary writers added to this a deconstruction of the human/animal dualism, arguing that there is not a morally-relevant difference between all animals.

Twentieth century pro-vegetarian writings expanded on all previous notions of kinship, sympathy, and a virtuous character to include animal rights, environmentalism, and feminism (Walters and Portmess 1999). Perhaps the most ardent vegetarian and animal rights spokesperson of the turn of the century was British teacher and social reformer Henry Salt (1851-1939). In his Humanitarian League memoir *Seventy Years among Savages* (Salt 1921), he challenged the notion of the English calling themselves *civilized* while practicing needless violence toward nature and human and nonhuman animals. Salt recognized that some animals and insects were killed in harvesting plant crops, but he distinguished between this harm being a necessity and the raising of animals for slaughter being unnecessary. Like Tolstoy and Kingsford, Salt spoke of vegetarianism as foundational to a virtuous life (Walters and Portmess 1999).

While Salt believed in the health benefits of vegetarianism, offering the working class populations as evidence, he argued that a humane motive should be the chief reason to go vegetarian, "as the moral basis of vegetarianism is the one that sustains the rest" (in Walters and Portmess 1999, 124) and creates a lasting commitment. Mohandas Gandhi (1869-1948) echoed Salt's assessment that vegetarianism should be motivated by morality, both to increase personal satisfaction and sustained commitment and to improve one's spiritual faculties. Gandhi stated:

> I found that a selfish basis would not serve the purpose of taking a [man] higher and higher along the paths of evolution. What was required was an altruistic purpose. I found also that health was by no means the monopoly of vegetarians. (in Walters and Portmess 1999, 142)

Philosopher Tom Regan (1975) emphasized that humans should specifically avoid *killing* animals, not just causing pain. Regan stated it did not matter if humans killed a nonhuman on an idyllic farm, on a factory farm, or in the woods, it was morally inconsistent to take away his/her life when one would not have taken the life of a human. This view relied on Regan's belief that there is no morally relevant aspect of humanity that separates it from other beings with a conscious interest in living. Since killing is to be avoided, Regan (2003) contended that the "total abolition of commercial animal agriculture" (1) is a goal of the animal rights movement.

Thus animal rights argues against so-called "humane" farming as a moral solution, as the vast majority of commercial animal farming involves killing at some point when an animal's production becomes unprofitable. I rarely hear of aging hens or cows being granted retirement, able to live out their full life on a commercial farm. Henry Salt (1921) noted that killing nonhuman animals (even painlessly) still constitutes a major harm because it cuts off their future opportunities for satisfaction and denies them the experience of enjoying their full life. In support of anyone's right to flourish, philosopher Gary Steiner (2008) contended that we "quite unnecessarily violate the prerogatives of animals to life and flourishing when we kill and eat them" (131). Even beyond killing, farming itself is a problem as it relies on domestication. Steiner bolstered the anti-domestication standpoint of the animal rights movement by saying "full recognition of the moral status of animals requires that we cease the practice of domesticating animals altogether" (130).

Peter Singer (1990) made a utilitarian argument against domestication when he stated that animal agribusiness, whether free range or intensive, is a speciesist practice because it controls and sacrifices the lives of farmed animals (major interests for the nonhuman animals) to satisfy humans' taste for flesh, milk, and eggs (minor interests for humans). Considering the fact that humans can healthfully live on a plant-based diet, a fact supported by the Academy of Nutrition and Dietetics (2009), Singer asserted that humans should make it a "simple general principle to avoid killing animals for food except when it is necessary for survival" (Singer 1990, 229). Philosopher Stephen Clark also theorized that because flesh-eating is largely unnecessary, it is morally untenable. Clark argued that it is hypocritical for philosophers to claim they are against unnecessary suffering, yet allow for it in diet, stating "those who still eat flesh when they could do otherwise have no claim to be serious moralists" (in Walters and Portmess 1999, 203).

Eating animals is not a trivial moral inconsistency, but rather a major, conscious inconsistency, according to philosopher Michael Allen Fox (1999). It represents a lack of accountability for the harm caused, whereas vegetarianism is "taking responsibility for a larger sphere of personal actions" (47). Vegetarianism should be conceived as an ethical philosophy in itself rather than limiting its connotation just to diet. It is for this reason that Fox titled his book *Deep Vegetarianism* – drawing a parallel with deep ecology's emphasis on making major changes and sacrifices (rather than shallow self-serving ones) for the good of the natural world. Fox said that vegetarianism "allows and even encourages philosophical and emotional connections between oneself and the world to develop" (36). With a similar focus on moral accountability, vegan wellness advocate Kathy Freston (2011) titled her popular book *Veganist* – a term she defined as someone who moves toward a plant-based diet based on a close examination into all the implications of their food choices.

Meat, Patriarchy, and Oppression

Carol Adams (1990) provided an important ecofeminist perspective on promoting vegetarianism. Adams connected patriarchy with the unjust domination of human women and nonhumans who are farmed and hunted. She asserted, "women and animals are similarly positioned in a patriarchal world, as objects rather than subjects" (168). Women and farmed animals both endure a "cycle of objectification, fragmentation, and consumption" (47). Farmed animals enter this cycle by being enslaved, butchered, dismembered, and consumed. Their body parts are sold in packages at the store quite fragmented and removed in appearance from the living, feeling creatures they once were. Adams concluded that, "eating animals acts as a mirror and representation of patriarchal values. Meat-eating is the reinscription of male power at every meal" (187).

The term "anthroparchy" (33) was coined by Erika Cudworth (2008) to describe the mutual roles of capitalist domination of nature and patriarchal oppression, as found in the meat industry. In her study of the British meat industry, Cudworth concluded:

> The object of domination in the manufacture of meat is patriarchally constituted. As such animals are largely female and are usually feminized in terms of their treatment. Farmers disproportionately breed female animals so they can maximize profit via the manipulation of reproduction. Female animals that have been used for breeding can be seen to incur the most severe physical violences within the system, particularly at slaughter. Female and feminized animals are bred, incarcerated, raped, killed and cut into pieces, and this tale of becoming-meat is very much a story of commodification. (43)

The sexist and speciesist connection between meat and masculinity has historical roots. In the book *Beyond Beef*, Jeremy Rifkin (1992) traced this connection between meat and masculinity back to ancient Egypt where the first universal religion was bull worship, based on the bull god Apis who represented strength, virility, and a masculine passion for war and subjugation. To mark the year's end, the Apis bull would be ritually sacrificed and fed to the king so he could incorporate the bull's fierce strength and power. More recently, in American culture, cowboys tamed the "Wild West" and turned it into a vast cattle grazing area, forever associating red meat with this brave and tough category of American men (Rifkin 1992). Adams (1990) highlighted men's traditional role in hunting animals and its perceived value in society: "Meat was a valuable economic commodity; those who controlled this commodity achieved power" (34). Adams referenced several anthropologists, such as Leakey, Lewin, and Sanday, who found that women's status was lower in societies where meat was important. Conversely, when economies relied on plant food, women held more status, and the society tended to be egalitarian.

While common knowledge may support the view that domestication of animals aided the advancement and progress of human civilization, sociologist David Nibert (2013) blamed oppression of nonhuman animals, via domestication of large animals for food, as the primary historical cause of oppression and harm toward *humans*. Both are motivated by elites' quest for material gain. Domestication is not a natural or "benign partnership" (12) but involves violence and objectification. It reduces the very existence of many species to only one recognized by their socially-constructed position as exploitable food objects. Because it desecrates their minds and bodies, domestication is referred to as "domesecration" by Nibert: "domesecration is the systemic practice of violence in which social animals are enslaved and biologically manipulated resulting in their objectification, subordination, and oppression" (12). He advocated for an end to domesecration as a necessary precursor to "development of a just and peaceful world" (7) because it "violates the sanctity of life" (12) for sentient beings.

Modern society views domesticated animals as "biotechnologies," according to sociologist Richard Twine (2010), particularly in animal agribusiness where corporations genetically design animal bodies as profitable protein factories. Twine built upon Barbara Noske's (1989) notion of the "animal industrial complex" to describe the vast embedded network of government and corporate relations that not only support an animal-based food culture but cultivate and preserve it (where industry is even more of a driver than farmers and consumers). This industrial complex is facilitating a "livestock revolution" where biotechnology seeks to transcend natural limits of production and to make Western diets and farmed animal breeds the norm in countries deemed to be underdeveloped and undercapitalized (Twine 2010). And this expansion is happening in an era of environmental crisis necessitating a reduction in the production of animal products (FAO 2006). Instead, the animal industrial complex is attempting to green the livestock revolution to be more efficient or produce energy from animal waste, thereby finding new justifications for continued or adapted forms of violence (J. L. Clark 2012; Twine 2010).

Naturalness Arguments Regarding Predation and Agriculture

Carol Adams claimed Western society maintains a social construction of humans as a natural predator and obligate omnivore so the necessity of our flesh-eating habit is not questioned. Under the omnivorous paradigm, vegetarianism is erroneously made to seem a naïve and feminized position that is ignorant of the laws of nature and anatomy. For example, food writer Michael Pollan (2006) accused vegetarian advocates of showing contempt for nature and predation, ironically demonstrating vegetarians' discomfort with acknowledging human's position as an animal. While it is true that many animal activists, along with many other people, are uncomfortable witnessing or partaking in the suffering and violence of predation, I do not assume that means that they do not understand the need for predation in nature or that they fail to embrace humans' animality in other ways. Humans can choose to define themselves largely as herbivorous great apes. In his book *The Unnatural Order*, Jim Mason (1997) cited anthropological theories that proposed for the first 25,000 years of Homo sapiens existence, prior to hunting and agriculture, they were largely herbivorous, as is natural to great apes. As Homo sapiens have migrated into ecosystems that do not provide adequate plant protein, they have the ability to become more omnivorous if necessary for adaptation. So while human history reveals periods of both greater herbivorousness and of greater omnivorousness, what is certainly unnatural is the vast consumption of animal products that has come to symbolize the normal American diet of the last half-century.

In opposition to Pollan's assertion that vegetarians fail to embrace their animality, philosopher David Wood (2004) argued that humans consume other animals to demonstrate control over the animal within themselves. Humans may surmise that the external animals they eat stand for the internal animal they must overcome. Wood contended that humans use meat-eating to continually reassure themselves of their powerful position in nature: "Might not the legitimacy of meat-eating rest, albeit precariously, not on our clear superiority to 'the animal' but on our need to demonstrate this over and over again?"

(138). In his popular book *Eating Animals,* Jonathan Safran Foer (2009) suggested that humans forget animal lives to avoid feeling shame over eating them. While we might recognize the similarities between us and the animals we eat, we "deny that these animal similarities matter, and thus equally deny important parts of our humanity. What we forget about animals we begin to forget about ourselves" (37). And whatever we want to forget about ourselves is relegated to our "animal nature" (37). One reason that psychologist Melanie Joy (2010) said that it is a challenge for humans to witness and confront our dissonance over eating fellow animals is that:

> If we no longer feel entitled to kill and consume animals, our identity as human beings comes into question. Witnessing compels us to view ourselves as strands in the web of life, rather than standing at the apex of the so-called food chain. Witnessing challenges our sense of human superiority. (143)

This notion of challenging identity and humanity's sense of entitlement reminds me that Donna Maurer (1995) had described pro-vegetarian rhetoric as being about (1) *entitlement*, as in nonhumans being entitled to live, and (2) *endangerment*, as in meat endangering animals and the planet. Yet, I think these two concepts of entitlement and endangerment also form the basis for most carnistic[5] rhetoric. While vegetarians may take an altruistic or social view of entitlement and endangerment as concepts defending the rights of others, carnists may see these concepts selfishly or individually as defending their own rights. I would say that carnists perceive of themselves as (1) *entitled* to eat animals, as humans are at the "top of the food chain" and are more important/significant than any other species, and (2) *endangered* by the threat of animal rights taking away their consumer right to sustain themselves on meat, eggs, and dairy, which they perceive as life-sustaining and strengthening.

Endangerment and vulnerability can also be linked to the human animal's own subconscious insecurity over being preyed upon and eaten, explaining Western society's insistence on killing other animals even in an age of plentiful food alternatives. This is part of Lee Hall's (2006a) theory that humans eat animals to demonstrate power over nature and maintain a self image as strong predators, not as weak prey.

---

[5] Here I refer to Melanie Joy's (2010) definition of *carnism* as a belief system justifying the human practice of eating certain nonhuman animals.

In agreement with Hall, Wood, Foer, and Joy, I think humans feel insecure if they don't see themselves as the top predator because a loss of exceptionalism makes them vulnerable to admitting they are just another animal among many on the planet. This lack of distinction means it could be justifiable for humans to sometimes be mere prey for carnivores, the way we now justify that role as natural for the animals whom we see eaten in wildlife documentaries and whom we farm and hunt ourselves. If you perceive, even unconsciously, of meat-eating as sustaining your life and privileged, secure status, it does explain the deep-seated psychological rationales for self-preservation that work to ensure that we humans do not become vulnerable to nature and carnivorous animals. If we don't see ourselves as an animal, we can stay *above* nature (and its perils), not be at the mercy of it.

Most who claim meat-eating is natural for the human animal fail to address how *unnatural* it is for any animal to breed and enslave others as a food source via *agriculture*. This implies that the human practice of hunting wild animals, a practice approximately 20,000 years old (Mason 1997) is more ethical under natural standards of predation than is agriculture, the latter being a more cultural domain. While agriculture might be largely unnatural to the animal kingdom, Pollan (2006) argued that farming animals is part of a natural evolution for human animals in symbiotic alliance with certain other animal species. But even within human evolutionary history, domestication of other animals for agriculture is a newer practice for the human species, originating approximately only 11,000 years ago (Mason 1997). As Singer and Mason (2006) noted, agriculture and other dominating practices allow humans to flourish but often in ecologically unsustainable ways (Singer & Mason, 2006). I contend that agriculture is less indicative of natural evolutionary adaptation and more a site of evolutionary role reversal, when the human species started to dominate and adapt nature to fit our own needs.

Utilitarian Arguments about Reducing the Most Suffering and Death

Even plant-based agriculture causes animals to suffer and die by displacing wildlife and frequently killing some field animals (including insects) in harvesting. However, because foraging would not sustain the human population at this point, plant-based agriculture is still largely required. And because a vegan diet is more sustainable and efficient at feeding people, veganism requires less land and kills fewer field animals than the current animal-laden diet (Singer and Mason 2006). Steven Davis (2003), an animal scientist, argued that it would be more humane for humans to eat grass-fed cows than to eat only plants, as that beef diet would actually cause the least animals to die. However, Gaverick Matheny (2003) refuted the validity of Davis's findings based on a miscalculation Davis made by assuming an acre of land could feed the same amount of people, when a plant-based diet could feed ten times as many. Therefore, Matheny calculated that a vegan diet kills one fifth as many animals as a diet based on grass-fed cows. Because a deontological or non-consequential ethic assesses motivation in addition to the rightness of the actions themselves, I argue that it is less of an ethical breach to *inadvertently* kill a number of wild nonhumans indirectly in necessary plant agriculture than it is to continue to legislate the intentional yet unnecessary breeding, captivity, and slaughtering of other animals (such as cows fed on grass).[6] The intentional farming of animals, in any form, facilitates an instrumental worldview that supports the subjugation of other animals, reducing them to just a means to an end.

Some have made the utilitarian argument that nonhuman animals have a better life and a less painful death living on a free-range farm than living in the wild (Pollan 2006; Sagoff 1993). I believe this argument is fallacious for a variety of reasons. First, it assumes that these farmed animals would willingly trade their lives and their freedom for the short-lived "security" of a captive existence prior to a guaranteed, premature slaughter by a human predator. Second, it implies that farmers save each domesticated nonhuman from an unsafe life in the wild, when the farmers actually created those nonhuman lives and, thus, are responsible for additional deaths. Plus, one of the ways farmers protect their "livestock" from nonhuman predators is to have the U.S. Department of Agriculture's Wildlife Services division kill millions of wild animals annually (Singer and Mason 2006).

---

6 Ideally the goal would be to develop harvesting practices that kill no one. This could be a new goal of agribusiness research. As it stands now, no consumer labeling exists (and very little information) about which types of crops kill more or fewer animals in growing and harvesting. This makes it nearly impossible for conscientious consumers to try to make purchasing decisions to avoid this kind of incidental violence.

Third, it suggests that all animal activists naively want these domesticated animals to be set free in the wild, where most are ill-equipped (due to years of selective breeding for heavy flesh, milk or egg production) to adequately take care of themselves and escape predators. Animal rights scholars admitted that ideally humans would not subjugate any nonhumans by domesticating them into a life of forced captivity and dependence (L. Hall 2006a; Regan 2003); however, these scholars did not suggest that existing domesticated farmed animals simply be set free, as that would be irresponsible and likely cause increased suffering. They simply suggested that we humans discontinue breeding other animals for our own purposes.

Natural and Cultural Ethical Guidelines for Meat-Eating

Although environmental philosopher Holmes Rolston (1993) defined the human practice of eating animals as a natural event that is subject to the laws of nature, not a cultural event subject to human ethics, I believe the variety of perspectives on the issue reveal eating is both natural and cultural. This reiterates my earlier contention that the bifurcation of nature/culture is somewhat necessary in determining ethical actions that affect both domains. I believe if the eating of others becomes necessary for the survival of a human in a certain place or situation, then flesh-eating, especially through low-tech hunting, becomes more justifiable under the ethical guidelines of nature. This places humans as another omnivorous animal in nature struggling to survive through hunting and foraging. Culture-based ethics requires that the moral agent be in a position to make a choice from free will. Therefore, when a human has a choice about what to eat, eating becomes more cultural and subject to human ethical systems rather than purely natural systems. This makes the routine choice to kill or exploit another animal for his/her body parts morally untenable, particularly when we enslave them for this purpose through agriculture. In agriculture, animals have no hope for escape or freedom from predation.

Other pro-vegetarian scholars bolster my argument in favor of some justification for humans eating meat when it is absolutely necessary for survival. For example, Gary Steiner stated this exception to the vegetarian rule "in those few instance in which ethics and survival come into conflict – as, for example, when an indigenous population simply cannot survive without using and perhaps killing animals – the imperative to survive should take precedence" (Steiner 2008, 132). Michael Allen Fox gave the examples of the Canadian Inuit or people thrust into extreme isolation due to disaster and said that humans, like any other animal, have the right to avoid "self-extinction" (Fox 1999, 160).

Because the ethicality of the human practice of raising and eating other animals has been debated since Ancient times, yet the practice is more prolific in Western culture than ever, it is perhaps not surprising that many meat-eaters are uncomfortable discussing farmed animal welfare and animal rights. Animal agribusiness capitalizes on this discomfort by communicating to the public using ambiguous discourse that largely hides the problems with animal production and seeks to increase its perceived ethicality, thereby reducing consumer guilt.

The Need for Deception in Communicating about Meat-Eating

In order to sustain the industrialized violence against nonhumans, such as in factory farming, Jacques Derrida admitted it is necessary for humans to "organize on a global scale the forgetting or misunderstanding of this violence" (Derrida 2002, 394). Animal agribusiness orchestrates this forgetting with consumers' implicit permission. Both Singer and Mason (2006) and Pollan (2006) discussed the difficulties they faced in trying to gain access to large animal farms for purposes of research for their books. Pollan observed, "the meat industry understands that the more people know about what happens on the kill floor, the less meat they're likely to eat" (304) as the conditions are "nightmarish" (318). And Singer & Mason noted that the media are often denied access, particularly with cameras, as agribusiness is resistant to allow any visual evidence of its practices to reach the public. In fact, Peter Cheeke (2003), an agricultural scholar, admitted that it is an ethical issue for animal agribusiness to purposely hide its industrial practices out of shame or fear that consumers will go vegetarian if they were to see the poor welfare conditions. This explains why many pro-vegetarian authors since the 18th century felt the need to vividly describe the cruelties of the slaughterhouse in an attempt to raise the public's awareness of the cruelty behind closed doors, even before the advent of factory farming post WWII (Walters and Portmess 1999). Pollan concluded that all industrial agribusiness and its consumers participate in a "journey of forgetting that could hardly be more costly, not only in terms of the animal's pain but in our pleasure too. But forgetting, or not knowing the first place, is what the industrial food chain is all about" (Pollan 2006, 10).

Most animal products, especially meat, are packaged with little information about the production conditions, facilitating no communication between farmer and consumer. Pollan (2006) found it ironic that consumers purchase food, something so fundamental to their health, based mainly on price. He posited that ignorance and cheapness are mutually reinforcing at keeping consumers apathetic to production issues and quality. But he noted a new trend in marketing animal products, such as cage-free or free-range, that includes explanations of improved animal welfare and wholesomeness. Pollan described the stories on the product packages as a new literary genre called "supermarket pastoral" (137). In the case of Petaluma's "Rosie, the organic free-range chicken," Pollan felt the marketing was fraudulent because, during a visit, he did not find the chickens roaming free and could not locate the idyllic pasture pictured on the label.

These new marketing stories rarely represent authentic communication about actual farm conditions but rather construct idealized visions of the kind of wholesomeness that consumers desire from farming (Pollan 2006; Singer and Mason 2006). It is just a newer tactic in the meat industry's history of using euphemisms to disguise unappealing or harsh practices (Adams 1990; Dunayer 2001; Glenn 2004). For example, industry and government officials refer to animals using marketing or commodified terms, such as *beef* instead of *cow flesh*, and *grain-and roughage-consuming units* or *inventory* instead of *animals* (Glenn 2004). Adams (1990) noted how terminology describing meat objectifies nonhumans and is purposely constructed via absent referents that allow consumers to distance themselves from their accountability in killing a living being. Describing the term *meat,* Adams stated, "something we do to animals has become instead something that is a part of animals' nature, and we lose consideration of our role entirely" (in Walters and Portmess 1999, 251). Through analysis of advertising and media images, Adams' (2003) book *The Pornography of Meat* exposed the hegemonic nature of patriarchy as an insidious force in American culture that allows inequality, such as that toward farmed animals, to virtually "disappear as a privilege and is experienced as 'desire,' as 'appetite,' as 'pleasure'" (171).

The public is complicit in the use of euphemisms to hide the unpleasantness of farmed animal slaughter, as Westerners seem to require self-deception regarding their unjust treatment of other animals in order to maintain their self-image as a civilized society. Linguist Jill Jepson (2008) noted that people use ambiguous terms for killing to avoid discomfort: "human beings avoid confronting the fact that they regularly kill living beings for the convenience and benefit of humans" (144). And almost a century ago, Henry Salt recognized that euphemisms were an impediment to becoming a peaceful society:

> The distinction between savagery and civilization is a matter of names... to use flattering titles as a veil for cruel practices gives permanence to evils that otherwise would not be permitted. Our present self-satisfaction in what we are pleased to call our civilization is a very serious obstacle to improvement. (Salt 1921, 239)

This sentiment is echoed by Derrida's prediction that industrialized violence against animals will have to change as it increasingly becomes a "spectacle" (Derrida and Roudinesco 2004, 71), forcing humans to decide that they cannot face the negative image this abuse creates of *themselves*. Besides the lack of public awareness about modern farming, the largest impediment to Derrida's prediction coming true is, perhaps, the continued restriction of meat discourse to the realm of farmed animal *welfare* instead of debating veganism and the rights of animals not to be farmed and killed (Freeman 2009a; Irvin 2007).

Yet, the public knows that farmed animals *are* killed to be eaten, and the common descriptor for this, *slaughter,* isn't exactly a euphemism. While modern agribusiness seems to prefer the euphemistic term *process*, I notice the public and media tend to bluntly use the traditional farming term *slaughter*. Jill Jepson's (2008) linguistic analysis revealed a human/animal dichotomy in terms of how the word slaughter is deployed. When used to describe the routine practice of farmers killing farmed animals, *slaughter* is a non-emotional term that merely describes an acceptable, legal industry practice of procuring a food resource. The farmer is not blamed for being cold or indifferent about killing, as it is his/her job. But when the term is used to describe humans killing each other, *slaughter* becomes an emotional term of outrage describing a brutal injustice, a criminal massacre. Jepson explained:

> Humans feel no need to differentiate between sanctioned and condemned forms of killing [nonhumans] because the taking of animal life is not generally a violation of morality or law… nor do humans feel the need for language that implicitly carries ethical evaluation. (143)

In terms of what kind of killer counts as a murderer, I note that if you use the term *butcher* to describe someone's paid position killing farmed animals, it benignly describes a legitimate profession serving the public's food needs. But if you call a human a *butcher* for killing any other type of animal (particularly a human), you have just accused him/her of being a violent and brutal murderer – a knife-wielding menace to society. But as long as we legally raise certain animals to be our food, killing them will not be seen as violent or criminal, rather it will merely be seen as professional.

In explaining how otherwise smart, caring people can support the butchering and consumption of animals, Fox (1999) theorized that people "compartmentalize" (40) farmed animals to distance the animal from the meat they are eating and also to justify their harsh treatment of one category of animal ("food") versus another (companions or wildlife). Similarly, in discussing America's irrational moral inconsistency in eating various animals, vegetarian author Mark Braunstein (2010) stated "inconsistency carried to the extreme of incoherency is the only excuse for feeling disgusted about the eating of horse flesh but not by the eating of cow flesh" (149).

This compartmentalized way of perceiving farmed animals is the focus of psychologist Melanie Joy's (2010) book dedicated to explaining the psychic numbing that people experience to justify their own behaviors (ex: eating animals) that are incongruent with their beliefs (ex: avoiding cruelty to animals). This defensive distortion of reality that keeps us from identifying and empathizing with farmed animals is enabled through a cognitive trio she lists as the "objectification, deindividualization, and dichotomization" (117) of farmed animals.[7] Farmed animals are presented as objects and property, not subjects.

---

[7] This is consistent with my prior findings on how the American news media primarily represented farmed animals: as bodies not beings, en masse not as individuals, and lacking the unique and respected emotional perspective that humans have (Freeman 2009a). I recommended that journalism: introduce us to the animals whose meat they present in the dining stories, share a farmed animal's perspective on his/her own situation, allow us to confront the violence of the slaughtering process, and debate humans' right to use/farm others.

They are perceived as a generic group without unique traits and personalities assigned to individual animals. Joy asked us:

> Imagine how you would feel if your package of hot dogs included a label with the name, picture, and description of the pig from whom the meat was procured, or if you became acquainted with one of the pigs who was to become your food. (119)

And society puts farmed animals into oppositional categories against other animals, such as edible, dumb, and ugly instead of inedible, intelligent, or cute. Joy noted that, "We retain false assumptions about the animals we eat so that we can continue to classify them as edible" (123). This cognitive dissonance process is supported by a psychological study finding that meat-eaters resolve the "meat paradox" (of liking animals and also liking meat-eating) by denying that nonhuman animals in general, especially farmed animals, have much capacity to suffer and are morally valuable: "When eating meat, people appear to suppress their moral concern, and this leads to a reduction in the perceived capacity of meat animals to suffer" (Loughnan, Haslam, and Bastian 2010, 159).

Meat-eating is more than a behavior, according to Joy (2010); it is a violent, mainstream, accepted *ideology* into which most of us are born and raised and generally taught not to question. She refers to this entrenched ideology as *carnism*. One of the reasons that carnism continues as a taken-for-granted truth is because it is justified through perpetuating the myth of the three Ns: "eating meat is *normal, natural,* and *necessary*" (96). "The mythmakers are the institutions that form the pillars of the system" (98) such as the rational and respected professionals in medicine, religion, education, industry, politics, law, and the media (the latter two being the most influential). These "moderate" professionals legitimize the system by talking about improving it (ex: farmed animal welfare) and by pathologizing anyone who significantly challenges the system (ex: veganism or abolition) as an "irrational extremist" (98).

The recent penchant for "happy meat," locavorism (eating locally raised animals), and cave-man diets is what Joy (2011) coined as *neocarnism*. This "carnistic backlash" (para. 3) is a shift from denial of a problem to a justification for certain meat-eating. It is a defensive reaction to the vegan and animal rights movements, revealing that people do see some validity in the three pillars of vegan rationales: animal ethics, environmentalism, and health. Joy proposed that neocarnism comes in three forms that address these three pillars and are also variations on the normal, natural, and necessary themes of carnism: (1) compassionate carnism (meat-eating is *normal* and can support "humane" farming), (2) ecocarnism (selective meat-eating is *natural* and sustainable), and (3) biocarnism (meat, eggs, and dairy are *necessary* for health).

Ultimately, neocarnism is just an adjustment to the same carnistic mindset without deeply addressing the core "irreconcilable conflict between caring about and harming other beings," (para. 14) particularly beings they have grown to see as food (Joy 2011). She gave these examples:

> Most proponents of compassionate carnism would not advocate slaughtering a perfectly healthy six-month-old golden retriever simply because her thighs taste good, just as ecocarnism proponents do not suggest consuming locally bred and slaughtered horses. And proponents of biocarnism insist not on the nutritional necessity of all meat but only of meat procured from 'edible' animals, such as pigs, chickens, cows, and fish. (para. 15)

As a psychologist, she characterized these neocarnistic justifications as rigid, reactive and defensive rather than flexible, open, and reflective. I find this interesting, as people usually characterize veganism as a rigid and inflexible extreme, yet Joy has turned the tables on this by framing neocarnism as an inflexible insistence on continuing to raise and eat farmed animal species. This fits with rhetorical scholars' suggestions that communicators make their notions flexible, adaptable, and progressive while making their opponents' ideas seem rigid and outdated (Perelman and Olbrechts-Tyteca 1969).

Joy (2010) suggested that the solution for carnistic denials and neocarnistic justifications is for society to acknowledge their own resistance to seeing farmed animals for who they really are, not just as food. People must have the courage to go against the social norm by witnessing the truth of carnism, even though it may cause them some emotional discomfort. Joy (2011) advised vegan advocates to enter into productive and compassionate dialogue with neocarnists, as the latter have demonstrated a promising shift in their consciousness about food, which is a sign of progress.

American Attitudes Toward Farmed Animals Today

Public opinion polls reveal that most Americans, approximately 70%, are generally supportive of the anti-cruelty goals of the broader "animal protection" movement, at least in comparison to a variety of other social movements (Humane Research Council 2012). Yet those same polls revealed that most Americans do not frequently talk about nor hear people talking about animal protection; in fact, a quarter of those polled said they had not talked about nor heard anyone talking about animal cruelty in the last three months.[8] This may be due to the fact that either they like to think that animals are being well cared for or that they do not actually care as much about nonhuman animals as they publicly proclaim.

Siding with the latter explanation, survey researchers at the University of Oklahoma's agriculture school concluded that people exaggerate their concern for nonhumans animals in surveys to seem caring, which researchers refer to as social desirability bias (Norwood, Lusk, and Prickett 2007). These researchers based this theory off the fact that their typical interviewee claimed that he/she personally cares more about animals than he/she thinks other people do or that he/she would be willing to pay more for humanely-raised animal products but other people would not. This indicates a public perception that farmed animal welfare is a low social priority and therefore will not exert much peer pressure. These agricultural economics researchers found that nonhuman animal welfare ranks much lower in importance than human welfare issues. In fact, the researchers' finding that Americans rank the financial wellbeing of farmers slightly higher than welfare of farmed animals allows the researchers to conclude that the agribusiness industry might defeat some farmed animal welfare reform measures by advertising that humane reform would be an economic hardship on farmers.

Based on all the comparisons between humans and "animals" in the survey, I conclude that the University of Oklahoma survey research (Norwood, Lusk, and Prickett 2007), which was funded by an industry source, seeks to ensure that the human/animal dualism and a speciesist attitude remains culturally intact. For example, they asked specific survey questions that enabled them to estimate that the American pub-

---

[8] These people must not have been in a conversation with me in the last three months.

lic believes the life of a single human being is equal to the lives of 11,500 farmed animals, in terms of how many farmed animals can suffer before their suffering potentially become more important than a single human's suffering.[9]

Yet while people may rank the *importance* of suffering differently, 80% of people said the *experience* of suffering was the same between humans and farmed animals in terms of ability to feel pain, indicating that people *do* know that farmed animals can suffer (Norwood, Lusk, and Prickett 2007).[10] And more than half the interviewees understood that farmed animals likely *are* suffering, as they believe that food companies put profits before welfare. Perhaps that is why many of these same survey respondents were in favor of stricter animal welfare laws and thought the government should actively promote farmed animal welfare.

Survey findings sometimes contradict each other because the Humane Research Council's (2012) surveys found there is a majority of people (at least 60%) who feel that laws protecting farmed animals in the United States are "adequate." But approximately a quarter of the population admitted that they don't really know much about these laws. An earlier telephone poll (Zogby 2003) found that approximately two-thirds of the U.S. population was unaware that farmed animals lacked basic legal protection in the United States, with over one third of those people mistakenly believing that state and federal anti-cruelty laws and the federal Animal Welfare Act ensured farmed animal welfare. This public misunderstanding about farmed animal protection

---

[9] These questions assessing comparative attitudes towards humans versus nonhuman animals indicates the industry might be putting the same emphasis on the importance of the human/animal dualism as I do, but based on different motives. I find it morally bankrupt to assess and quantify how little someone's life is worth – how much people discriminate against farmed animals – so that you can continue to profit off of that discrimination. While I find these survey questions offensive, they reveal to me that the industry and I agree that continued use of nonhuman animals is reliant upon the public maintaining the hegemonic belief that humans are wholly different and more important than mere "animals."

[10] I think this public awareness of farmed animal sentience is important considering Joy's and Loughnan's contention that people tend to deny the suffering of farmed animals so that they can feel comfortable eating them and the fact that the animal protection movement feels the need to keep educating people that farmed animals are suffering. The public already knows this on many levels, but I think they are not asked/pressured to make farmed animals a priority.

may account for the fact that over 70% of people polled in 2003 believed farmed animals are treated fairly in the United States.

But what does "welfare" or "humane treatment" mean to people? The University of Oklahoma researchers found that basic food and water and vet care ranked the highest, followed by ability for the animals to be outdoors and behave naturally (Norwood, Lusk, and Prickett 2007). "Freedom to behave naturally" ranked higher than just being safe and comfortable indoors. Both animal protection researchers and agribusiness researchers try to determine whom the public finds credible on defining welfare standards. Animal protection researchers found that veterinarians rank the highest, and, similarly, agribusiness researchers found that scientists are often deemed credible. This public trust in science seems to be promising for the industry, as the agriculture school researchers suggested that if scientists could prove that animals had lower stress levels and were more protected in crates than in larger pens or outdoors, then the public would be more supportive of the confinement practices (Norwood, Lusk, and Prickett 2007). The report also indicated that researchers are concerned that the public makes naïve moral decisions when they automatically dislike crates, as consumers just don't understand the supposedly legitimate or benign reasons the industry strictly confines many farmed animals. When it comes to quantifying industry practices as "humane" to please consumers, it seems safer and more manageable for the agribusiness industry to capitalize on the supposed "objective" credibility of science rather than rely on the subjective and uneducated moral judgments of the public. I presume that research grant money will be funded in pursuit of these goals for science (especially in schools of agriculture) to validate profitable industry practices as "humane."[11]

The last section of this chapter expands on current food practices, as it provides a brief overview of the status of farmed animal produc-

---

[11] This University of Oklahoma survey report aimed at the industry was disturbing to me, as it seemed to be asking questions of consumers to better determine how agribusiness can continue to operate business as usual and ascertain how much more money they can charge if the public or government requires higher welfare standards. I didn't get the impression the industry was genuinely concerned about improving animal welfare or reducing suffering of animals but, rather, was more concerned about how little change they could get away with while still placating an increasingly conscientious consumer market.

tion, meat consumption, and vegetarianism today, especially in the United States.

HUMAN EATING HABITS

Meat Consumption Rates

As human population has doubled between 1961 and 2008, the world's annual production of animal flesh has quadrupled, rising from 156 billion pounds to 617 billion pounds (Cohen 2010). That equates to an average human being consuming 92 pounds of animal flesh per year in 2008. But people in "developing" countries consume less than average – about 68 pounds per person – while people in "developed" countries consume 183 pounds per person. This will vary by country, as a 2007 meat consumption study found that the lowest rates of annual per capita animal consumption were in India and Bangladesh (less than 8 pounds per person), and many nations on the African continent were under 54 pounds per capita, as were countries such as Peru, Indonesia, Turkey, Pakistan, Haiti, and North Korea (Barclay 2012). Many European countries and South American countries were in the range of 150 – 200 pounds per person. Some populous nations that fall in the middle range of 100 – 150 pounds were China, Russia, Japan, South Korea, Saudi Arabia, Mexico, Malaysia, and South Africa. But the highest rates of animal consumption per capita were in "Western" countries –Luxembourg topped the charts at over 300 pounds per person, Australia and the United States tied for second at 270 pounds per person, and countries such as New Zealand, Spain, and Austria also came in high – falling between 227 – 257 pounds per person (Barclay 2012).

If we look at total meat consumption just in the United States, it has drastically increased over the last century (going from around 10 billion total pounds in 1912 to 52 billion pounds in 2012) (Larsen 2012). As monumental as 52 billion pounds is, this latest number represents a slight decrease in meat consumption from the peak high in 2007 at around 55 billion pounds (Barclay 2012). The cow has always been the most popular land animal consumed in the United States, reaching an all time high in the 1970's and decreasing steadily since. But just recently, Americans have started to consume more chickens than cows, with the rise in the consumption of chickens increasing quite steadily since the 1950's. Next come pigs, the consumption of whom has fluctuated a bit but is currently similar to the rates from 100 years ago. The next most popular land animal that Americans consume is the turkey, the consumption of whom has steadily increased over the last century but remains much lower overall compared to the other species mentioned above (Barclay 2012).

In total, the number of *land* animals that U.S. slaughterhouse workers killed in 2009 was at least nine billion. While overwhelming, the only good news is that this number represents about 300,000 fewer lives taken than in 2006[12] (Humane Research Council 2011). The number of *sea* animals that humans kill for food is indeterminate, as their lives are measured by weight not individually, but it is estimated that Americans consume 17 billion sea animals annually. These statistics represent land and sea animals whose bodies ended up in the marketplace. But each year hundreds of millions of additional animal lives go unaccounted for, as agribusiness kills them or leaves them to die on the farm, in hatcheries, in transport, or as "bycatch" animals discarded in the sea by the fishing industry (Singer and Mason 2006, 112).

Vegetarianism Today

It is likely that hundreds of millions of people worldwide eat a vegetarian diet. Determining exact rates of vegetarianism is challenging based on confusion over terminology and inaccuracy in reporting (it is assumed people exaggerate their meatlessness). A variety of polls from different countries estimate that, globally, India has the highest rate of vegetarians, with potentially 35 - 40% of Indians eating no animal flesh, thanks to the popularity of Hinduism and Buddhism (Wikipedia 2012). Many European nations (excluding Spain and Portugal) have a fairly high and growing rate of vegetarianism, varying between 3 – 10%, especially countries such as the UK, Germany, Italy, the Netherlands, and Switzerland. Similar rates may be found in Brazil and Taiwan. Canadian rates of vegetarianism may be similar to those in the United States.

Approximately 7% of the U.S. population self-identified as vegetarian on polls taken earlier in the 2000's, yet pollsters estimated that

[12] Between 2006 and 2009 the number of animals killed spiked up before decreasing again.

the percent of the population who was *actually* vegetarian was likely between 2 and 3%; and about 1% of these people, or approximately a million Americans, were vegan (Maurer 2002; Singer and Mason 2006). More recent polls show a slight increase in vegetarianism, as the percent of Americans who now say they never eat any type of meat is 5 to 6% (Newport 2012; Stahler 2011), and about half of this population also said they don't eat dairy or eggs either, which equates to 2.5% of Americans eating a vegan diet (Stahler 2011). On a recent Gallup poll (Newport 2012) 2% of respondents self identified as "vegan" (Note: the survey didn't provide a definition for the term). The Vegetarian Resource Group poll (Stahler 2011) conservatively estimated the actual percentage of Americans who are vegetarian is likely still between 2 and 3%, and they quantified that, considering the lowest and highest percentage numbers from polls, this equates to between 5 to 12 million vegetarians in the U.S. today. In a hopeful trend, they estimated that an additional third of Americans (while not 100% vegetarian) are eating meatless meals "a significant amount of the time" (para. 2).

The typical person attracted to vegetarianism is a young, white, middle-class, atheist female (Maurer 2002), and the Gallup poll found that single people are twice as likely as married people to be vegetarian (Newport 2012). Similar rates of vegetarianism were found throughout all regions across the United States and the split between genders was fairly even, but women were more likely than men to eat meatless meals more frequently (Stahler 2011).

The main reasons people say they go vegetarian is for health and/or ethics, with environmental sustainability being another motivation (Maurer 2002). People who go vegetarian for ethical reasons tend to be more committed to remaining vegetarian. Because health-motivated vegetarians may be tempted by the convenience of a meat-based diet and new lower-fat meat items, Maurer (2002) posited, "promoting concern for animals and the environment is essential to the advancement of the vegetarian movement" (45). This coincides with Henry Salt's and Gandhi's belief that vegetarianism should be promoted on the rationale of ethics more so than health, based on both the deontological belief that the ethical rationale was more solid and the utilitarian belief that it created greater long-term commitment to vegetarianism (Walters and Portmess 1999).

CONCLUSION

This chapter helps to define and bolster the animal rights and vegetarian moral philosophies that should serve as a basis for informing the messages of animal rights organizations studied in this book. The next chapter focuses on communication theories that can guide activists in making communication decisions, and it includes deontological and utilitarian framing debates among animal activists and scholars about how to construct campaign messages designed specifically to protect the lives of farmed animals.

Chapter 3

Activist Communication Strategy & Debates

This chapter provides an overview of scholarly literature on strategic communication, particularly on behalf of social movements, getting more specific to rhetorical options for animal rights and vegetarianism at the end. I begin with a basic overview of the social constructionist perspective on communication and meaning-making. It elucidates the signifying power of language to serve as a site of ideological struggle in determining what society sees as common sense truths. This notion of communication's intentionality helps explain what I mean when throughout the book I say some entity "constructed" a certain meaning (rather than just saying they "re-presented" it). I then provide some ethical guidelines for persuasive communicators, including ethical challenges especially pertinent to radical social movements. Next I examine literature specific to communication of social movements. This includes the special challenges that social movement organizations face in designing persuasive communication campaigns, conveying counter-hegemonic ideas to the public, forming an identity, attracting productive media attention, and utilizing visuals in the digital age.

Because this book specifically examines vegan message framing, a large section of this communication chapter is dedicated to framing theory, frame resonance, and the frame alignment process, mainly drawing upon social movement theory from sociology. To help examine the framing debates within social movements over deciding how radical or conservative to be, examples are included of internal framing debates within the American women's rights and abolitionist movements of the 19th century. This leads into a similar discussion of the current animal rights movement's ideological framing debates over whether to promote animal rights or welfare. This debate then narrows to explore animal activists' deliberations in food campaign frames over whether to promote veganism exclusively or to also promote industry welfare reforms. The chapter closes with an overview of scholarly studies specific to the framing of vegetarianism.

Communication and the Social Construction of Reality

Communication is the heart of culture; it is the process through which a shared culture is created, modified, and transformed (Carey 1989) – a cyclical process where ideological meaning is produced/encoded, consumed/decoded, and then reproduced/transformed into social practice (S. Hall 1980). This process is not neutral; critical media scholar Stuart Hall (1982) proposed that the dominant elite, using tools like the media, manufacture consent by shaping and reinforcing hegemonic ideologies so they seem like taken-for-granted common sense. However, he believed audiences are also active subjects who have some ability to resist intended meanings, for example with queer readings of heterosexual characters (S. Hall 1980).

A cultural view of communication sees language as a purposeful social construction of meaning, based on the politics and values of society, and not merely as a reflection of some biological or empirical truth (S. Hall 1997). For example, in binaries, such as *mind/body*, *culture/nature,* or *human/animal*, meaning is situated through opposition, according to hierarchical values that rely upon or even exaggerate notions of difference, so that we know something by what it is not (Chandler 2002). Jakobson introduced the idea that in each binary one term is marked and one is unmarked; the unmarked term, such as *human* or *mind*, is more fundamental, naturalized, and prioritized while the marked term, such as *animal* or *body*, relates to it as secondary, lacking, or deviant (Chandler 2002). Derrida's (1976) work on deconstruction seeks to challenge the power inherent in prioritizing one binary term over its supposed opposite by revealing the logical inconsistencies and instabilities of these classic oppositions.

French philosopher Michel Foucault's notion of discourse adds to the theory that language itself is embedded with naturalized power by showing how discourse, or a system of representation, has the power to rule in or out ways of talking about a topic (S. Hall 1997). Foucault (1990a) conceived of discourse as the historically-influenced construction of knowledge (ideas, images and concepts) that is used to organize thoughts and action on a topic in a certain culture at a certain point in time. Discourse has the capacity to control social practices and preserve institutional power through its management of what is considered to be "truth" and knowledge in a society (Foucault 1980) .

However, it is more important to ask *how* something becomes true rather than what "is" true. For Foucault (1980), truth is produced "only by virtue of multiple forms of constraint" (130). He conceived of discourse as working like a grammar to determine the conditions for what *could* sensibly be uttered but not dictating the exact utterances themselves (Hoy 1981). Because of its social influence, discourse "is the thing for which and by which there is struggle. Discourse is the power which is to be seized" (Foucault 1990b, 1155).

Social Movement Communication Challenges

Social movements often seek to directly challenge the dominant discourse and its hegemonic construction of reality. They must convince the public that not only is the public's accepted view of reality based on a faulty premise but the situation deserves to be defined as a "problem" that warrants their immediate attention (Stewart, Smith, and Denton 1984). DeLuca & Peeples (2002) refer to this aspiration as "making the mundane malevolent" (145). The communicative task of social movements is to enable target audiences to perceive problems as severe and unresolved by authorities, creating a sense of urgency that motivates social intervention. Activist tactics include the use of persuasive words, stories, gory pictures, and revelations of inconsistencies in institutional practices (Stewart, Smith, and Denton 1984).

The "public sphere" in which social movements operate cannot be optimistically defined as a civic forum for public consensus but, instead, as a corporate-owned site of structured ideological negotiation (Eley 1992). On this uneven playing field, and with fewer resources to gain support, social movements face pressure to use socially acceptable language to be taken seriously as credible and reasonable. But it is hard to appeal to values that are part of the very system that they are challenging (Cox 2006). In deciding how critical to be of the status quo, Todd Gitlin (2003) noted that social movements must walk a fine line between extremes; if they are too moderate, they risk being assimilated and "blunted" (290), yet if they are too critical, they risk being marginalized and trivialized. Robert Cox (2006) explained that radical social movements are often constrained to use less critical discourse that appears "reasonable" and stays within "symbolic legitimacy boundaries" (61) as determined by the hegemony of mainstream discourse, making it hard to achieve ideological transformations of the status quo. I like Foucault's (2000) suggestion that discursive transformations necessitate criticism of the status quo:

> Criticism (and radical criticism) is utterly indispensable for any transformation. For a transformation that would remain within the same mode of thought, a transformation that would only be a certain way of better adjusting the same thought to the reality of things, would only be a superficial transformation. (457)

Related to this debate over how critical or "radical" activist messages should be, is the decision whether to base their persuasive appeals on the public's individual self-interest or on appeals to altruism and social justice (Cox 2006). Similarly, it relates to the question of how large of a change activists should request from the public. These questions are particularly relevant to movements that ask for change on behalf of other species. To produce wider acceptance and to seem reasonable, there is a pragmatism to just asking people for easier or smaller behavioral changes (Meatless Mondays or carpooling once a week) and hoping that will eventually lead them to adopt the needed larger changes, as they begin to identify with that pro-social behavior and what it says about them (Cooney 2011). But change strategist Tom Crompton (2008) argued against aspects of this "foot-in-the-door" spillover strategy on the grounds that it isn't producing enough of the critical environmental changes needed and that, when it is based off of appealing to self-interested motives, it can promote values that are counter-productive to environmentalism and caring for others. Rather than appealing to people's finances or self-interest, appeals based on concern for the environment are more likely to lead to spillover into other pro-environmental behaviors. Similarly, Neil Evernden (1986) argued that altruistic, non-anthropocentric appeals are necessary to win long-term support for the environment because appeals to the public's self-interest are ultimately just unproductive short-term strategies that reinforce a view of nature as a resource. When it comes to "bigger than self" issues, like species extinction or factory farming, appeals to self-interest and extrinsic values (social status, money, or rewards) are less motivating than are common-interest and intrinsic values (community, personal growth, benevolence) (Crompton and Kasser 2009). Even in our consumer culture, this research asks us to have faith in people's depth and core values, as even people who are extrinsically-motivated can have their natural inclination for intrinsic values engaged if they are asked to reflect on what is really important (Chilton et al. 2012).[1]

Cox (2006) distinguished between the pragmatism and short-term focus of *campaign* rhetoric and the long-term goals of *critical* rhetoric. The former often benefit from self-interested, reasonable appeals, while the latter are meant to more broadly challenge existing values to envision new worldviews. The problem I see is that activist organizations may prioritize short-term campaigns, largely excluding long-term critical rhetoric so that its existence in the public discourse is marginal, seeming all the more radical. In this book I propose a blending of critical rhetoric into everyday campaigns.

Sociologist James Jasper (1997) did not prefer to think of activists as just another type of public relations campaigner or pragmatic marketer; he perceived of them as artists who serve a unique pro-social role as moral visionaries and innovators, opening up mental possibilities for cultural self-reflection and improved ways of being. They balance out the prominent efficiency and profit-oriented discourse of corporations and the state. Similarly, Deluca & Peeples (2002) said that radical activists' image events are "visual philosophical-rhetorical fragments, mind bombs that expand the universe of thinkable thoughts" (144).

And in today's globalized digital media environment *visual* rhetoric, in particular, plays an essential role in creating presence for one's issues through the eyes of international audiences who do not all speak the same languages (Jamieson 1988). DeLuca & Peeples (2002) concluded that the Habermasian notion of the dialogue-driven public sphere is now more akin to a "public screen" (127), where media spectacle and image events constitute a visual version of discourse and

---

[1] While I don't ascribe to a behavioral approach to social change, especially the notion of encouraging small incremental individual behaviors, I still draw upon some psychology studies, particularly those focused on changing values. In discussing how to facilitate values-based ideological change, for purposes of this book, I primarily utilize communication scholarship related to discourse and framing, particularly in social movement studies. Another useful approach to examine are sociological theories of social practice that address change at a fundamental socio-political level, including political economic critiques (see Shove 2010; Webb 2012).

sense-making that construct our reality, as images themselves *are* "a central mode of public discourse" (144). Shocking images have proved useful in the recruitment of activists to the animal rights movement (Jasper 1997). Therefore, image events, such as the use of graphic visuals by activists, should not be interpreted merely as desperate stunts but more as a communicative and democratic necessity (DeLuca 1999; Jamieson 1988).

The news media offer limited opportunities for successful delivery of confrontational rhetoric for social movements. As a result, PETA now prioritizes more pop-cultural, sometimes comical, approaches to reaching audiences through entertainment media, even though this tactic runs the risk of seeming sensational, trendy, trivial, or even offensive (Simonson 2001). Ingrid Newkirk, PETA's founder and director, explained that PETA is forced to turn to more sensational campaigns and be "stunt queens" (Younge 2006, 12) to get any attention for issues because the news media do not find everyday animal exploitation newsworthy. This approach has faced critique when racialized or sexualized human bodies are used to gain media attention for animal issues in ways that can be perceived as objectifying or exploitative (using, and potentially furthering discrimination toward, one marginalized group to help another) (Adams 1995; Deckha 2008; Kim 2011; Pace 2005). While seemingly productive in some short-term utilitarian sense, it is ultimately counterproductive if it furthers objectification and discrimination of any sentient being, as it is those mindsets that the animal rights movement is ultimately trying to overcome. But in her feminist analysis of PETA campaigns, Deckha (2008) cautioned that each use of female bodies in animal campaigns should be considered in its discursive context, rather than being condemned outright as sexist, as some may be "productive and subversive" (38) of species boundaries and resistive to oppressive systems.

Another challenge the animal rights movement faces in media campaigns is redefining normal animal use as abuse:

> Animal movement activists seek to stigmatize and mark as deviant what many people perceive as normal, legitimate, mainstream activities...The animal movement must transform the moral meanings associated with the worst of these practices, redefining them as socially irresponsible." (Munro 1999, 36)

Munro (1999) agreed with Peter Singer's (1990) contention that the animal rights movement's survival depends on its ability to maintain the moral high-ground in campaigns. The opposition knows this, as a common strategy of animal use industries is to frame animal rights activists as misanthropic (Munro 1999). This tactic serves to call into question the movement's morals, or at least their moral priorities, playing off of the public's deep-seated beliefs in human superiority. I believe this is another example of how addressing the false human/animal dualism is foundational to any long-term change toward a less speciesist society.

Advocacy Communication Ethics

As activist groups are passionate about their cause and may appear close-minded or biased, their advocacy communication is sometimes derided as manipulative "propaganda." However, propaganda has some distinctive negative characteristics that distinguish it from mere persuasive speech. Propagandists prioritize ends over means, putting their own interests above the audience's, which may involve purposely distorting or misrepresenting information, sometimes even concealing the source of the message from the audience (Jowett and O'Donnell 1999). Therefore, ethical advocacy communicators must avoid the kind of manipulative, misleading, and reductionist message constructions that are characteristic of propaganda, such as: reliance on authority figures; use of unverifiable abstractions; belief in a fixed, polarized world; reduction of complex issues into simplified cause and effect; and emphasis on conflict over cooperation (Black 2001). Advocacy organizations are not required, however, to provide a full summation of all the facts and various sides on an issue, as they are not journalists. But, in support of truth, they should seek to provide enough context to "genuinely inform" audiences instead of creating "false impressions" (Martinson 1996, 44).

Persuasive communication does have ethical guidelines, which Baker and Martinson (2001) represented as five main principles (using the acronym TARES): Truthfulness of the message, Authenticity of the persuader, Respect for the audience, Equity/fairness of the appeal to the audience, and Social responsibility for the common good. In an article I published in the *Journal of Mass Media Ethics*, I argued that activist communication is distinct enough from corporate, nonprofit, and governmental communication to warrant its own ethical guidelines, based on the challenges social movements face in redefining entrenched social practices into social problems (Freeman 2009b). In my recommended communication guidelines, the TARES principles of truth, equity, social responsibility, and, to a large extent, authenticity, remain applicable to social movement persuasion. This would prevent activists from including false information, intentionally misleading people, targeting vulnerable groups manipulatively, and communicating socially-irresponsible messages (such as promoting violence, racism, or sexism).

But if a counter-hegemonic social movement organization is seeking social justice and peace, then I make an exception, of sorts, for their communications when it comes to the TARES principle of respecting the audience. I advocate that activists may criticize and offend (disrespect) some of the audience's[2] discriminatory *values* and problematic causal *behaviors* (but not to disrespect them as morally-relevant *beings).* Specifically I mean activists should be allowed to make a candid, and potentially disrespectful, critique of mainstream audience values when it conflicts with the duty to promote social justice, in this case privileging overdue respect for marginalized victims more so than respect for audience feelings. For example, consider a controversial vegetarian campaign such as PETA's "Holocaust on Your Plate" campaign. If utilitarian calculations were allowed to compare the amount of harm caused to the meat-eating audience (in this case emotional harm/offense)[3] with the amount of harm meat-eaters and animal agribusiness cause to PETA's primary moral claimants – the farmed animals (in this case severe emotional and physical harm, resulting in death) – it could be seen as justifiable to create some short-term emotional dissonance in meat-eaters while attempting to mitigate the greater harm these consumers cause other animals. In

---

[2] It becomes more acceptable if the potentially disrespectful messages are aimed at culpable or powerful audiences rather than more "innocent" or irrelevant parties, and it should keep social equity and social responsibility principles in mind so as not to further discriminate against historically marginalized groups.

[3] Although in this case, those with ties to the Holocaust tragedy may experience more or a different type of emotional distress than the average meat-eater. But intention is important in ethics, and PETA was not trying to offend or belittle WWII Holocaust victims, but rather to show sympathy for all victims of mass murder. Their intention was to criticize government and industries that oppress and kill (and thereby get the public to end complicity and withdraw their support). For more detail, see my conference paper on PETA's campaign (Freeman 2007) as well as Kim's (2011) analysis on this campaign.

essence, social movement organizations need the moral flexibility to sometimes be controversial, shocking, and impolite in pursuit of the greater good. This is especially relevant since a social movement's version of what is "good" may not be appreciated by mainstream society at that point in time, so viewers may critique both the communication means and the end goals as harmful.

Additionally, while I contend that truth, a core communication principle, remains largely applicable to all organizations, its counterpart, authenticity, may require some consideration for activists. Communication authenticity is related to the character of the communicator, as demonstrated in virtues such as integrity, loyalty to stated causes and institutions, and sincerity and genuineness (Baker and Martinson 2001). But to increase resonance and acceptance with mainstream audiences, counter-hegemonic social movement organizations are pressured to be less ideologically-authentic – less critical, candid, and progressive – so it is understandable if they adapt to fit the primary motivations of their target audience in order to achieve campaign goals. But I argue that ideologically-authentic communication not only has the ethical advantage of being more open and representative of the activist organization's actual agenda and motivational values, but also may be more effective at building a morally-rational foundation for the long-term change they seek (Freeman 2009b). Even if most social movement campaigns fail to result in victory (at least in a short-term sense), Jasper (1997) claimed they still have social value because the protestors' importance "lies more in their moral visions than their practical accomplishments" (379).

The next section examines social movements specifically, drawing largely upon sociological literature. It begins with an overview of the communication practice of framing, especially for social movements, and ends with a discussion of frames used by animal rights organizations.

STRATEGIES FOR SOCIAL MOVEMENT ORGANIZATIONS

Framing Overview

To be effective, social movement organizations must not only mobilize resources and capitalize on political opportunities, they must strategically use communication to awaken cultural consciousness and create a collective identity for change. This cultural turn in social movement theory arose in the mid 1980's, with a focus on strategic framing of ideas (McAdam, McCarthy, and Zald 1996). Framing theory largely developed from Erving Goffman's (1974) conclusion that people use expectations and interpretative schemas, called *frames*, to make sense of all situations in life. Humans must mentally frame their everyday experiences to be able to cognitively comprehend and manage their reality and make decisions about appropriate actions. These frames serve to make things meaningful, organize experiences, and guide actions.

Many communication scholars have applied Goffman's ideas on framing to explain how communicators package their ideas through frames and how others respond to these frames. For example, Robert Entman (1993) acknowledged the power of framing to identify problems and solutions: "to frame is to select some aspects of a perceived reality and make them more salient in a communicating text, in such a way as to promote a particular problem definition, causal interpretation, moral evaluation, and/or treatment recommendation" (52). And Todd Gitlin (2003) described the importance of framing to meaning-making by stating: "Frames are principles of selection, emphasis, and presentation composed of little tacit theories about what exists, what happens, and what matters" (6).

Social Movement Collective Action Frames

Because the purpose of social movement communication is to build mass support for a campaign to mobilize target publics into action, activists engage in what is known as *collective action framing* (Snow and Benford 1988). Sociologist William Gamson (1992) defined the three components of collective action frames as: (1) injustice (a problem exists and it is important), (2) agency (assurance that we can fix it if we work together), and (3) identity (side with us). David Snow & Robert Benford (1988) more generally defined the three core tasks of framing as: (1) diagnostic (define the problem and possibly attribute blame), (2) prognostic (define solutions), and (3) motivational (encourage collective action).

The practice of selecting the *diagnostic,* or problem, component of the collective action frame can be contentious within a movement (Snow and Benford 1988); disagreements may occur not only in defining the problem for the public but also in assigning blame, as causality for problems is often multi-faceted and complex. An activist organization's diagnostic and prognostic frames should align, as the definition of the problem constrains the range of pertinent solutions (Benford 1987). The *prognostic,* or solution, component of collective action frames is often influenced by external factors that may create a need for the organization to counter-frame remedies offered by their opponents and rationalize recommendations (Benford 1987).

The *motivational* component of collective action frames must construct a compelling rationale that serves as an inspiration to engage in collective action toward the proposed solution (Benford and Snow 2000). To garner this support, motivational frames often rely upon an appeal to shared values and identity, demonstrating alignment between the goals of the organization and those of the target audience. To achieve this, activists can practice frame alignment processes, discussed below, such as frame amplification, frame extension, and frame transformation (Snow et al. 1986).

By tapping into existing social values or beliefs, frame *amplification* is the clarification of an interpretive frame so that the frame bears on a particular issue and people see the connection (Snow et al. 1986). All movements utilize frame amplification, but it is particularly useful to movements whose values contradict society's core values and are in need of greater support (Berbrier 1998). Frame amplification involves amplifying and clarifiying values and beliefs. *Values* refer to guiding behaviors or states of existence that society deems worthy of protection and promotion (Rokeach 1973). As values exist in a hierarchy that vary by individual, activists must elevate a presumed value to create salience for it in the mind of the viewer and demonstrate its direct relevance to the issue at hand (Snow et al. 1986). *Beliefs* describe relationships and are "ideational elements that cognitively support or impede action in pursuit of desired values" (470). Frame amplification must address the following core beliefs affecting desire to participate in collective action, such as people's need to believe: the problem is serious, certain parties are to blame, change can happen if they act collectively, and their assistance is necessary and socially acceptable (Snow et al. 1986).

Frame *extension* is produced by extending the boundaries of a social movement's framework to show it includes other causes and issues that are important to a group of potential adherents (Snow et al.

1986). This is useful for creating coalitions with other social justice groups. To be ethical, activists need to be sincere and avoid using frame extension to merely gain additional resources. Another caution is that extending the number of issues one advocates for can backfire by diluting the specificity of the activist organization's original cause, resulting in increased disputes among core supporters.

The last alignment process, frame *transformation*, consists of creating new meanings and values, often by changing old meanings (Snow et al. 1986). Frame transformation could be characterized as *ideological* transformation (Oliver and Johnston 2005). It is particularly necessary when the values promoted by a social movement, such as animal rights, do not resonate or may even appear antithetical to conventional lifestyles (Snow et al. 1986). New values must be planted in society and erroneous beliefs reframed, such as a change in the way a domain of life is framed so that what previously seemed acceptable is reframed as unjust or problematic. This can sometimes be done under a broad or global interpretive frame transformation, such as a metanarrative of peace, which reframes many domains of life under a new universe of discourse.

Framing is an integral part of ideological transformation, but Pamela Oliver and Hank Johnston (2005) clarified that frames and ideology are not identical. Ideology is a broader ethical belief system that informs frames, so that frames can be seen as reductionist presentation strategies for the ideology. Similar to Foucauldian discourse, ideologies serve as both a constraint and a resource to the framing process, and the resulting frames help scholars empirically observe ideology at work (Snow and Benford 2005). Frames are important because they serve as a recruiting tool for ideologies (Oliver and Johnston 2005).

Political strategist George Lakoff (2004) noted the centrality of framing to enacting social change, as change cannot occur without issues being strategically *re*framed:

> Reframing is changing the way the public sees the world. It is changing what counts as common sense. Because language activates frames, new language is required for new frames. Thinking differently requires speaking differently. (p. xv)

Varying Characteristics of Frames, such as Resonance

The characteristics of frames can differ between organizations within the same movement. Benford and Snow (2000) noted four main characteristic of frame variances: (1) problem identification and direction of attribution; (2) flexibility and rigidity, inclusivity and exclusivity; (3) variation in interpretive scope and influence; and (4) resonance. Regarding the latter, how much a frame resonates with its audience is primarily dependent on two things: *salience* with the values prioritized by the individual and society, and perceived *credibility* of the speaker and their message. Activists increase the salience of frames by highlighting values and beliefs that are central in the targeted individual's hierarchy of values and beliefs (Rokeach 1973). Salience also requires that frames be commensurate with and relate to the target's personal experiences by not seeming too abstract or distant from his or her everyday life (Benford and Snow 2000). And to increase credibility, activist messages should be congruent and logically consistent with the stated beliefs and actions of the activist organization. For example, organizations should avoid such contradictions as preaching nonviolence and compassion and then acting or speaking in an aggressive, rude, or violent manner, as was demonstrated by the radical fringe of the anti-abortion movement (Johnson 1997). As an effective example of both credibility and salience, the successes of the Southern Christian Leadership Coalition in the U.S. civil rights movement can be partially credited to the accessibility and resonance of their framing around popular American principles of Christianity and democracy (McAdam 1996). Additionally, there was congruence between the civil rights protesters' nonviolent resistance tactics and their frames of morality and justice.

Another framing variable to consider is that the more inclusive and flexible a frame is, the greater the opportunity it has to evolve into a master frame (Benford and Snow 2000). Master frames, such as rights, justice, and democracy, are the broadest in interpretive scope and can be used across different movements. A wider sense of unity and agreement can be created by appealing to abstract values (as opposed to concrete/specific concepts), as the ambiguity allows different people to connect with the concept via their own interpretations (McGee 1980; Perelman and Olbrechts-Tyteca 1969). And it is useful for marginalized groups to emphasize values that prioritize *quality*, such as rightness or uniqueness, over quantity, since they don't have numbers/popularity on their side. Organizations can create a sense of presence and connection for these abstract values by utilizing compelling visuals or mythic/familiar narratives to introduce individuals in a memorable story-telling format (Perelman and Olbrechts-Tyteca 1969; Polletta 2006). For greater acceptance, new ideas should connect with culturally-accepted and historically-situated ideas and narratives (Ryan 1991; Tarrow 1998), such as how animal rights organizations expand the democratic notion of human rights to extend to nonhuman animals. Rather than merely presenting facts and rational arguments, activists are encouraged to use a values-based appeal to create cultural resonance by framing themselves as a progressive group trying to right moral wrongs in a conflict, similar to other freedom fighters in history (Ryan 1991).

Similarly, linguist George Lakoff (2004) suggested that advocacy organizations avoid talking primarily in terms of policy, facts, or negations and, instead, talk in terms of a clear set of simple *values* that accurately reflect what the organization stands for and express its "moral vision" (74). For example, Lakoff concluded that U.S. conservative political groups were generally better at framing than progressive groups because conservatives "say what they idealistically believe" (20) while progressives take the utilitarian approach of relying on polls to decide how radical or moderate to be. He advised communicators to avoid using their opponent's language and frames when trying to negate them, as that only serves to inadvertently reinforce the opponent's way of seeing the world. In support of authentic communication, an organization's language and frames should fit and emphasize *its own* worldview and values.

Psychologists give us a pragmatic reason to follow this advice; repetition of an activist's position and terminology can help the public see those new ideas as more legitimate, as what sounds familiar will seem to represent the majority rather than minority. People, as social animals, are biased toward what seems popular, as popularity validates an idea or practice as more credible (Cooney 2011).

To better understand the communication challenges faced by key rights movements in creating moral progress in the United States, I will discuss internal framing debates within the early women's rights and civil rights movements. In pursuit of legal reform, they debated the extent to which their messages should criticize or challenge the status quo discriminatory (racist and sexist) ideology. This relates to the previous discussion of how social movements struggle to create resonance and alignment for their frames with the public on issues over which they may disagree. And movement leaders often disagree over these tactical framing decisions that appear to involve compromising ideological authenticity for public acceptance and tangible results.

IDEOLOGICAL FRAMING DEBATES IN U.S. SOCIAL MOVEMENTS

Framing is a competitive process where leaders in challenging movements struggle externally with counter-movements and authorities to define frames, but there are also *internal* framing struggles over tactics and goals (Zald 1996). This section's framing debates stem from the early stages of U.S. human rights movements for women's rights and civil rights in the 19th century. They serve as useful examples as they bear some relevance to today's animal rights movement framing struggles, since animal rights is also in a fairly early stage of development and seems radical for the time.

Framing of Women's Rights

In her book *Man Cannot Speak for Her*, Karlyn Kohrs Campbell (1989) claimed that women of the early 19th century began to acknowledge their own need for rights as they were denied the right to be spokespersons and leaders of any significance in the other social reform movements that they pioneered. They convened the first women's rights convention in 1848. The convention's Declaration of Sentiments, which was heavily ridiculed by the male-dominated press, based its manifesto on extension of natural rights to women. This reaffirmed American values of democracy and justice and promoted reform not revolution.

Campbell (1989) identified two major contradictory framing choices made by women suffragists, one was more *ideologically pure* and the other more *politically expedient.* The former frame was the more "radical" or oppositional argument drawn from the women's rights convention, stating that women deserved rights based on the equality guaranteed to citizens per the U.S. constitution. The other frame, political expediency, was a more pragmatic and moderate approach asking for the vote on the basis that it would be socially beneficial if women could spread their innate moral virtues to the public sphere and better facilitate their caretaking roles as wives and mothers. The latter approach was considered more feminine, as it was selfless and altruistic, and its message played off of stereotypes of female purity and domesticity. The former approach of asking for equal rights for oneself seemed more masculine and self-centered. The women's rights movement did experience factioning over sending these mixed messages that confused the identity of women's suffrage in the eyes of the public.

Campbell (1989) explained that movements must balance contradictory internal and external pressures to maintain buy-in of current members while still attracting new members to the cause. The paradox is that if you maintain ideological purity to advance a new worldview, you increase your *internal* unity through radical identity, and the conflict is then oriented externally toward the public. But if you use politically expedient ideologies that are less threatening to the status quo, it creates more *external* unity with the public but more disagreements and factions internally within the movement. The more moderate and feminine suffrage appeals "exemplify the seductive strategies that the oppressed are constrained to use when they lack the legal, political, and economic power to effect change" (96).

While Lucy Mott, Henry Blackwell, and Frances Willard used this more moderate and expedient approach, Elizabeth Cady Stanton and Susan B. Anthony were ideological purists, using a natural rights argument. When speaking to legislators, Stanton and Anthony were somewhat aggressive and used legal and democratic arguments. They pointed out contradictions in the American legal system, such as women being denied the right to be tried by a jury of their peers and women being taxed without being able to vote. Taxation without representation was a powerful argument that resonated with democratic struggles in American history (Campbell 1989).

Additionally, Anthony and Stanton connected women's rights with other classic struggles that resonated with most men at the time: the American revolution, the civil war, and the protestant revolution. Similarly, they used the analogy of comparing American men to tyrants such as kings, feudal barons, and popes – all very un-American traits that made democratic American men appear hypocritical. Later, Stanton chose an indirect attack by focusing on the victim and comparing women to enslaved persons instead of attacking the men as victimizers (Campbell 1989).

Framing of Civil Rights and the Abolition of Slavery

In his book *Forerunners of Black Power: The Rhetoric of Abolition*, Ernest Bormann (1971) identified two main rhetorical styles of abolitionist speakers in the mid 19th century, both influenced by Puritan preaching heritage: *agitation* (used by William Lloyd Garrison, Wendell Phillips, and Frederick Douglass) and *conversion* (used by Theodore Weld). While agitation and conversion are similar, respectively, to Campbell's (1989) dichotomized rhetorical styles of ideological purity and political expediency in the women's rights movement, Bormann's abolitionist's agitation style seems more radical and aggressive, in content and delivery, than the women's rights version, as the agitator rhetoric was more revolutionary than reformist in its critique of the United States.

Bormann (1971) analyzed these two rhetorical styles according to the two major challenges facing social movement leaders. First, leaders must mobilize people to their cause by showing them what is wrong and how they can fix it, similar to the diagnosis and prognosis components of collective action frames (Snow and Benford 1988). In doing so, the movement must historically situate its cause into the cultural narrative while creating its identity and vision (Bormann 1971). For example, the movement's vision may be a return to a better time or perhaps characterizing itself as part of an ongoing progression toward a better society. Second, similar to the identity component of collective action frames (Gamson 1992), Bormann (1971) said the movement must create a group identity to gain commitment from members by making itself meaningful to members' lives.

When it came to abolitionists mobilizing people and situating the movement in American culture, both agitators and conversionists used testimony and evidence from Southerners and former enslaved people to shock the public with the brutal facts of life under slavery (Bormann 1971). Both types of rhetoric viewed slavery as a sin, but conversionists more often used biblical rhetoric. Agitators focused on means, principles, and morals, and were openly critical of institutions and anyone supportive of slavery, calling Americans hypocrites. They historically situated themselves as new revolutionaries outside of the

American experience because they believed society was corrupted by racism. On the other hand, conversionists focused more on ends, as the goal was to be persuasive. They appealed to people's noble interests for the American dream and principles of freedom. They saw their movement as the better part of the American dream – of guaranteeing the natural rights of humanity. The conversionist approach was effective, and many new anti-slavery societies started because of it.

In addressing Bormann's (1971) second reform challenge of creating group unity, identity, and commitment, both agitators and conversionists sought to make the abolitionist movement a major force in their members' lives by labeling it as a moral duty. They both claimed it was a righteous cause sanctioned by God and was worth sacrificing and suffering for. Agitators were more stringent in their demand that slavery be outlawed immediately and full political rights be granted to all men. They did not water down the message to make it more appealing and felt disruption was necessary because it got attention and headlines. Conversionists, however, were more moderate at first and said abolition could happen gradually and black men should not be granted full voting rights immediately. But conversionists eventually came around to the full rights position, which Bormann argued was the rhetorically stronger position. Bormann (1971) described the conversionists as being more successful than agitators at reducing internal fighting and at converting more people and appealing to outsiders.

In the civil rights movement a century later, Dr. Martin Luther King's rhetoric is more similar to conversionists with his promotion of democratic American values (Bormann 1971). But it was firm like the agitators about the immediate need for equal rights. King's rhetorical style was evangelical and drew much support from churches. Contrastingly, the Black Power movement was more agitation-oriented and attacked cherished American values and social structures as inherently racist.

In conclusion, it seems to me that the preferred framing approach is a mix of the two styles. Social movement communicators should promote strong moral values and be uncompromising on rights like the agitators and ideological purists. But they should make sure to situate themselves as reformers within national cultural values, and possibly use a softer sell, to be more appealing and make progress like the conversionists and political expedients. Just as Campbell (1989) and Bormann (1971) distinguished between more and less critical, or ideological, framing approaches in historical rights campaigns, in the next section I examine similar framing debates in the modern day animal rights movement regarding whether to prioritize rights versus welfare, or, alternatively, abolition versus reform.

IDEOLOGICAL FRAMING DEBATES IN THE ANIMAL RIGHTS MOVEMENT

Distinctions between Animal Rights and Animal Welfare

Legal scholars Gary Francione (1996) and Lee Hall (2006a) suggested that animal rights activists should more authentically align their rights ideology with their activist strategies. Both authors drew a distinction between animal welfare and animal rights ideologies and favored the latter. Francione (1996) said animal *rights* is about justice and the abolition of animal exploitation and not allowing other animals to be treated as a means to human ends. A rights philosophy demands the "incremental eradication of the property status of animals" (4) to raise them to the level of "personhood" (6). He claimed:

> The rights advocate makes one thing very clear: that animal rights is a position of the *outsider* who ultimately seeks a paradigm shift in the way that law and social policy regard the status of animals, as well as in the human/animal relationship. (219)

Hall (2006a) defined animal rights as a deontological ethic granting nonhumans the right to privacy and freedom from human intrusion. It is an argument against use and domination in favor of freedom.

Conversely, these authors defined animal *welfare* as a mainstream philosophy that merely regulates animal exploitation and the suffering of the nonhuman animals we control. Francione (1996) stated that animal welfare has the following characteristics: (a) it recognizes animal sentience but believes they are not as worthy of moral respect as humans, (b) it recognizes the property status of nonhuman animals while wanting to limit the rights of property owners to be more humane, and (c) it accepts trading away the interests of nonhumans in favor of human interests only if the latter are deemed significant and necessary. Francione added that welfarists are more concerned with reducing the *suffering* of nonhuman animals used by humans rather than preventing the *use and death* of those animals (Francione and Garner 2010).

Francione (1996; Francione and Garner 2010) claimed that the modern day animal rights movement is largely a hybrid of both rights and welfare. Hall (2006a) also noted that few animal protection organizations actually promote rights; humane groups clearly promote welfare, and, ironically, even many radical direct action groups ultimately focus on welfare and suffering. Francione (1996) explained that many animal rights organizations operate on the belief that they must use a welfare platform to get to the eventual goal of rights. He argued that a welfare approach is "structurally defective" (4) at accomplishing an abolitionist rights agenda. It is "counterproductive on both theoretical and practical levels" (5), as a social movement must align its ideology, goals, and strategy for logical consistency. Francione (1996; Francione and Garner 2010) and Lawrence and Susan Finsen (1994) admitted that a largely welfarist animal protection movement has raised awareness of animal suffering over time, but it has not achieved the goal of decreasing the number of animals who are exploited.

Both Hall (2006a) and Francione (1996) critiqued utilitarian philosophies of animal *ethics*, like Singer's, as well as utilitarian activist *strategies* that fail to align the message and tactic with the kind of end world they seek. Favoring pragmatism, the animal rights movement fails to connect its practice with its theories on animal rights. Hall and Francione used metaphors such as *treadmills* and *chasing one's tail* to describe the futility of welfare reforms that seek to chip away at the myriad ways nonhuman animals suffer within an exploitative system. Any such victories are shallow, as they merely mitigate a few of the endless array of symptoms but do not get significantly closer to eliminating the *root cause* – an instrumental view of nonhuman animals as property. Francione recommended "we need to move the discussion to the issue of animal use rather than animal treatment" (Francione and Garner 2010, 213)

In support of Baker & Martinson's (2001) authenticity principle in advocacy communication ethics, I believe a rights message from a rights organization is honest communication that authentically represents the group's goals without hiding aspects that might be unpopular and less mainstream. Regarding being open about one's agenda, Francione (1996) stated:

> Although many animal rights organizations claim to embrace the complete abolition of animal exploitation as a long-term goal, they often couch this message in more "conservative" terms in order to make their message more acceptable to the public. The problem with this approach is that it allows animal exploiters to respond that animal advocates are not honest or that they have some "secret," agenda, which is arguably harmful to the overall credibility of the movement. (117)

Regarding a willingness to be candid, Hall (2006a) provided the example that most animal protection groups promote so-called "humane" farms instead of asking supporters to go vegetarian.[4] Hall lamented that more advocacy groups did not "cultivate a public demand for peaceable, animal-free farming unabashedly" (99) because it is defeatist and timid to give up on replacing exploitative systems and settle only for demanding improvements to the system. Hall likened this easy-sell approach to following a corporate marketing model that adjusts to fit the status quo and treats citizens like consumers by offering them a bevy of appealing choices. She argued that these expedient tactics just end up distorting the issue.

An additional advantage of maintaining an ideologically-based frame is that it enables the activists to control the discourse by defining the problem around the root cause of exploitation and enslavement instead of allowing the issue to be limited to animal husbandry/welfare or human self-interest. Regarding the weakness of using the self-interest frame, Francione (1996) explained:

> These arguments shift the moral focus from issues of justice for a disempowered group to the self-interest of the empowered group and open the debate to various empirical considerations, such as how dangerous meat eating really is or whether vivisection is really "scientific fraud." (118)

Regarding the weakness of a welfare frame, Francione argued that limiting the frame to welfare fits with a mainstream industry perspective, allowing industries to claim they are in agreement or compliance with humane treatment. This may inadvertently benefit animal exploitation industries, whose strategy has been to alienate the animal rights activists by labeling them as misanthropic and militant in favor of coopting the more conservative welfare groups and humane messages.

---

4 As readers will see in future chapters, the animal rights organizations that I examined for this book *did* advocate veganism as the ideal rather than free-range farming. But their focus on condemning *factory*-farming may imply there is an ethical way to farm animals rather than not domesticating them at all.

Therefore, using a rights campaign that questions the legitimacy and very existence of animal-use industries has the advantage that it cannot be co-opted by them.

Framing around Incremental Abolition Goals

In *The Animal Rights Debate*, Garner framed abolition in somewhat limited terms as consumer vegan campaigns – impractically crusading for morality instead of crusading for legal or policy changes (Francione and Garner 2010), but welfare or reform campaigns are also moral campaigns and vegan or abolitionist campaigns can also be focused on political or legal change. In advocating for abolition, Francione (1996) understood the pragmatic need for animal activists to feel effective and not campaign in vain for rights or seek violent revolution, so he acknowledged that rights, not just welfare, can be gained in stages of incremental abolition. But he cautioned that animal rights organizations should ensure their campaigns for incremental change are actually based on *rights* not welfare principles. He provided the following examples of rights campaigns: asking people to go vegan or to boycott companies who test on animals, ending the use of certain animals in certain kinds of exploitation, protesting or exercising civil disobedience against an exploitative industry, or banning certain hunting or experimentation practices or devices. While Francione mentioned that a rights agenda could include banning cruel farming practices like battery cages or dehorning, he cited a caveat by Robert Garner (1993) who argued that this could just lead to public support for less cruel animal farming, a concern that is only partially mitigated by including a vegan message in the campaign.

Francione (1996) also cautioned against relying too heavily on regulatory reforms of industry. The industries have the law on their side as they are owners of animal property, so until animals are not considered property, he said it is futile to request significant legal change when "the legal system structurally limits the scope of reform to what is dictated by the instrumentalist position" (171).

The next section narrows the debate over the use of animal rights versus animal welfare frames and applies it to the framing of farmed animal issues and vegetarianism specifically.

Farmed Animal Framing Debates in the Animal Rights Movement

For a little over a decade, most major animal rights organizations, and even some animal welfare groups such as the Humane Society of the United States (HSUS), have made farmed animals a primary focus, as farmed animals comprise the vast majority of nonhuman animals killed in the United States. Instead of primarily promoting veganism, the recent trend for some animal protection organizations is to encourage farmed animal welfare reforms, as has been successful in Europe. In some cases, the animal protection organizations promote less cruel farm products, such as cage-free eggs, in addition to vegetarianism. This shift toward farmed animal welfare reforms has sparked debate within the animal rights movement over effectiveness, authenticity, and integrity in movement strategy. In this section, I include arguments both for farming reform and for veganism.

In favor of welfare reform frames

Those activists who argue in favor of working with the meat industry to institute higher animal welfare standards often use utilitarian arguments about it being more effective both at eventually promoting veganism and currently reducing the amount of suffering billions of animals endure. Advocates for welfare reform argue that to insist only on veganism, when it is adopted at such a slow rate, is tantamount to the animal rights movement turning its back on the billions of animals who currently suffer. Instead, welfare advocate Robert Garner noted the strategic benefits of campaigning to eliminate whichever farming practices can be deemed as causing *unnecessary suffering* (of little benefit to humans), such as the British campaign to ban live exports of farmed animals (Francione and Garner 2010).

Singer (2006) admitted he has become more of an incrementalist since the reasonable arguments for veganism that he presented in *Animal Liberation* in 1975 have failed to make veganism mainstream. He now argues that raising awareness about the lack of farmed animal welfare in the United States will serve to raise public consciousness that minor improvements are still not enough. Miyun Park (2006), of the HSUS, took a pragmatic approach by arguing that welfare strategies attract more media attention to educate the public about poor farming conditions, as referendum efforts increase visibility for factory farming abuses. I agree that the more laws there are on the books to protect farmed animals, the more those laws will have a chance to be broken; this would make animal farming more newsworthy in a criminal context, although the focus remains on debating animal treatment not animal use. It would help if the news did not frame those proven instances of abuse as an independent case of corporate malfeasance – "one bad apple" spoiling an otherwise decent industry (Freeman 2009a).

When farmed animal welfare issues make the news (whether it be from a vegan-inspired undercover investigation of a slaughterhouse, or a reform-inspired statewide referendum to ban crates), this coverage can lead to temporary aggregate decreases in meat consumption. A Kansas State University study found that Americans purchased 2 - 5% less pig and chicken meat following media coverage of farmed animal welfare issues. This is presumably out of disgust or concern for how we raise and kill animals, as people didn't switch to different types of meat; they replaced or reduced animal flesh for quite a few months (Tonsor and Olynk 2010).

Economics can play a role in reducing animal product consumption as well. Park (2006) and Cooney (2012) provided the utilitarian motive that farming reforms often drive up prices, which would reduce consumption of animal products, as this has been documented in Europe. I think it is probably true, if less cruel methods were outlawed in the United States, that cost-conscious consumers would eat fewer of these domestic products (depending on how pricey they were), but one has to take into account that cheap, factory farmed animal products would likely still be readily available and popular as long as there are global free trade agreements. Due to globalization, it seems to me the best way to ensure supply decreases is to decrease demand for any animal products.

Farm Sanctuary Director Gene Baur (Bauston 2006) argued that the movement should not dichotomize welfare versus rights as both can be accomplished by asking the public to view farmed animals as sentient beings instead of commodities. Similarly seeing rights and welfare as unified, Peter Singer and Bruce Friedrich (2006) contended that the philosophical argument for granting animals freedom from cages also "logically demands that we cease to exploit them for our own ends" (12).

Being realistic about the slow growth of veganism and the "remote" (viii) chances of persuading everyone to stop eating animals, Singer (2008) is understandably concerned about the billions of animals that are and will continue to suffer on farms. So he called for a blended strategy:

> This is not an either/or choice. The animal movement should continue to promote a cruelty-free vegan lifestyle, and to encourage those who are not vegans to eat less meat and dairy products. Recognizing that not everyone is ready to make such changes, however, the movement should also be involved in improving the welfare of animals used in commercial farming. (viii)

He stated that the stakes are too high and "catastrophic" (viii) for the animals and the environment for the movement to confine itself to only promoting veganism. Additionally, welfare advocates cite a correlation between a nation having greater legal protection for farmed animals and that nation having greater numbers of vegetarians, with Britain being an example (Cooney 2012; Park 2006; Singer 2006). This indicates that reform and abolition can work in conjunction instead of at odds with each other.

In favor of primarily promoting vegan frames

Many activists do not believe the contention that animal agribusiness reform is philosophically consistent with animal rights (Dunayer 2006; L. Hall 2006b; Lama 2006; Mark 2006; Torres 2006). These activists emphasized that the purpose of animal rights is to promote life, freedom, and respect through the abolition of speciesist practices such as industrialized animal slaughter. They do not believe any implicit or explicit promotion of "happy meat" aligns with that life-affirming goal. Australian animal advocate Patty Mark (2006) encouraged animal rights activists to fulfill their unique purpose in global animal discourse by asking activists, "If *we* are not going to give the hard message for what the animals need, who is?" (25).

Some activists emphasized that working with animal agribusiness weakens the movement's integrity and credibility. Former cattle rancher turned vegan activist Howard Lyman (2006) maintained that if the meat industry is wrong and animal rights organizations team up with them, then the animal organizations are wrong too. Documentarian James LaVeck (2006b) claimed that financial incentives encourage both industry and animal rights organizations to negotiate the "price of public concern for animal suffering" (21), as both groups end up growing and getting the resources they need.[5] To dissuade animal organizations from helping animal agribusiness profit, LaVeck suggested:

> We don't need to be a part of dreaming up the details of the industry's new and improved systems of exploitation, and we certainly don't need to put our good names and our movement's credibility behind the questionable products that result. (23)

Many activists expressed concern that by promoting farmed animal reforms they were sending conflicted and mixed messages that weakened their position by revealing an identity crisis. Sociologist Bob Torres (2006) argued that it shows conflict, weakness, and defeatism to promote a kinder version of speciesism. Torres stated that welfare is untrue to animal rights ideology and turns activists into advocates for exploitation. LaVeck (2006b) agreed that animal rights messages need to be clear, strong, truthful, and morally consistent. If animal rights organizations negotiate with industry, it sends a complicit message that eating animals' meat is a necessary evil and all that activists and the public can realistically do is try to mitigate suffering. LaVeck claimed that through welfare reform, activists are introducing "moral ambiguity into situations where the boundaries between right and wrong must never be allowed to blur" (23). Francione (1996) noted that it can be hard to maintain a distinction between encouraging farm industry reform and implicitly promoting the resulting animal products.[6]

---

[5] LaVeck's website humanemyth.org provides examples of collaboration between animal protection groups (such as the HSUS, PETA, and Farm Sanctuary) and ag industry producers and retailers (ex: Wolfgang Puck, Niman Ranch, and Whole Foods).

[6] I found this to be true when I had our student animal rights group at Univ. of Oregon participate in the HSUS's campaign to get campuses to switch from battery cage eggs to free-range eggs. While students certainly did learn about the egregious factory-farming practices that our university was supporting, I was more concerned that many students apparently did not feel chickens were very worthy (of attention or higher egg prices). After the frustration of failing to even get this welfare improvement widely adopted, and not wanting to support the purchase of anything taken from an animal, our group's next campaign was a Vegan Challenge. That campaign was rhetorically much easier for me to construct and promote.

While Singer does believe activists should reform industry, in his book with Mason (2006) they noted that from a consumer perspective veganism is superior because it provides ethical clarity, making it easier to make food choices and stick with them; this contrasts with so-called "conscientious carnivores" who will always be plagued with the ethical dilemmas of both determining how humane is humane enough and whether or not the farms from which they buy are truly acting as humanely as they claim. If Singer and Mason's argument is turned around on the animal rights movement, it seems to imply activists should prioritize vegan campaigns, as they lack the ethical ambiguity dilemmas of promoting so-called "humane" farms.

The authenticity and power of animal rights communication is also compromised when animal rights organizations sanction agribusiness's appropriation of principled terms that guide the animal movement, such as *compassionate* and *humane*, thereby lowering the threshold for what these terms mean in society. These terms represent something positive to the public, but in reality the movement has allowed them to misleadingly represent something less positive – a softer version of killing and approved amounts of suffering (LaVeck 2006a).

Ethical communication issues with welfare reform frames

New York activist Eddie Lama (2006) indirectly addressed communication ethics by noting that humane reform activists think they are tricking the trade into eventual abolition when in reality the trick is on them, as the industry uses the activists as economic leverage to sell so-called "happy meat." This echoes Francione's (1996) concern that when animal rights campaigns claim to be reasonably reforming agribusiness to better fit mainstream animal welfare standards, the public may be mislead about the activists' more "radical" abolitionist agenda of moving them toward veganism. The industry can then point out this breach in communication ethics to undermine the credibility of animal rights activist groups.

Controlling the discourse by defining the problem around rights

All questions, strategies, and solutions stem from how the problem is defined. James LaVeck (2006a) and author Joan Dunayer (2006) suggested that animal rights activists retain control of the discourse by defining the problem as exploitation and slavery not husbandry and cruelty. A husbandry frame is too narrow, excluding both the larger issue of the nonhuman animal's right not to be exploited and humans' lack of need for their exploitation in the first place. This echoes Hall's (2006a) and Francione's (1996) suggestions that animal rights campaigns be brave enough to maintain a rights ideology, since being outside the mainstream is essential to transforming the status quo. The more animal activist campaigns focus on reforming, instead of transforming, the mainstream use of nonhumans for food, activists create framing challenges for themselves because a vegan solution then remains more radical in the eyes of the public. LaVeck (2006a) argued that when some animal rights groups work with industry on welfare reforms it can set back the whole animal rights movement from a framing standpoint:

> The focus of public dialogue irrevocably shifts from the questionable morality of using and killing animals, to an elaborate, endless wrangle over how the deed will be done – conditions, treatment, standards and regulation. In this new framework, public calls by animal advocates for the boycott of all animal products, for nonparticipation in exploitation, have no place. Such talk is now an embarrassment for the participating animal groups, and a joke for the meat industry people. Such talk is now relegated to the realm of "radicalism." (20)

Vegan frames and the connection with environmentalism

In considering whether to prioritize humane farming reforms or veganism, LaVeck (2006b) reminded activists that animal products, even ones that are cage-free, tend to be less sustainable and contribute to greater environmental devastation than plant-based foods (see Freeman 2010c). Unlike farming reform campaigns, vegan campaigns have the added benefit of being able to emphasize sustainability, which benefits all animals, including humans and wild/free nonhuman animals. Conversely, welfare campaigns are limited to focusing on a reduction of suffering for captive nonhumans and typically fail to relate to environmental issues.

Similarly, Hall (2006b) reminded animal rights activists to see the big picture of how their work connected with other social causes, such as environmentalism, in seeking to transform humans' dominionistic attitude toward others and nature. "We're seeing the biggest set of extinctions and the most ominous climate indicators in modern history," Hall stated. "Negotiating with industries is fiddling as Rome burns. We should be very busy learning a different way to think about other animals and the earth" (25). Hall's (2006a) belief that animal rights is about letting nonhumans live free from human interference requires that animal activists work toward ensuring a healthy environment with habitat for all animals. Hall advocated that animal rights organizations promote veganism to work toward this mutual goal of animal rights and environmentalism.

Meat-reduction

While American vegan advocates Matt Ball and Bruce Friedrich (2009) do promote veganism as the ideal, they advocated that animal activists promote "incremental evolution" (52) to veganism, by asking people to give up eating as many individual animals as they can. This means privileging small animals killed in higher numbers who don't provide as much food per life taken, such as sea animals and birds, then moving onto larger animals like pigs and cows. They think it's better to give consumers something besides an all-or-nothing choice (meat-eating or veganism) by praising any life-saving measures consumers take, even if they just stop eating chickens. Pragmatically, they noted that, "several people reducing their animal consumption by half saves as many animals as one person going vegetarian" (62).

Regarding the animal rights goal of veganism, I use the following section to explore Americans' attitudes about vegetarianism and meat-reduction as well as some scholarly studies examining framing of vegetarian messages.

Framing of Vegetarianism

Advocacy recommendations based on consumer opinion polls

To frame vegetarian messages effectively, it helps to understand public opinion and motivations regarding meat-eating. A study by the Humane Research Council (HRC) (2007) found that, while total vegetarianism remains a marginal diet in the United States, 13% of Americans consider themselves semi-vegetarians and over a quarter of the population says they are actively reducing their meat consumption.

This latter group, comprised largely of women and older consumers, is primarily motivated to reduce meat consumption based on self-interest, such as health, rather than on animal or environmental protection. While consumers view vegetarian foods, such as fruits, vegetables, and whole grains, as healthy, they also generally believe that some animal products, such as dairy, chicken, and fish, are also healthy.

Vegetarian advocates are challenged by survey findings that reveal 80% of Americans do not intend to ever fully eliminate meat from their diet, based on concerns that it may be unhealthy to do so and their overall preference for the taste of meat. However, almost a quarter of the population is interested in reducing their meat consumption by half. Therefore, the HRC (2007) study suggested that it would be more effective for vegetarian advocates to promote meat *reduction*, rather than vegetarianism:

> For an adult audience, meat reduction is clearly more acceptable than complete veg*ism, and there is strong evidence that this approach to veg*n advocacy would persuade more people. Moreover, there is evidence that those who start to reduce their meat consumption become more open to both further reduction and possible elimination of meat from their diet. (7)

Consumers interested in meat reduction and vegetarianism have concerns about its potential inconvenience and cost, so advocates should provide consumers with practical information on how to overcome these barriers.

When presenting rationales to the public, pro-vegetarian organizations should avoid mass advocacy campaigns in favor of more targeted approaches (HRC 2007). Consumers often need to hear a variety of reasons for going vegetarian, and it is likely most effective to lead with a health argument followed by a concern for animal suffering. While the health rationale, in particular, and the environmental rationale, to some degree, are more useful at encouraging people to *reduce* meat consumption, the animal suffering rationale is most effective at motivating people to *eliminate* meat. Therefore, the HRC (2007) concluded that it may be most strategic for different advocacy organizations to use different appeals, such as some concentrating on meat reduction for health reasons and others on vegetarianism for animal protection reasons.

Communication tactics of U.S. vegetarian advocates

Sociologist Donna Maurer (2002) studied vegetarianism as a movement promoted by animal protection organizations and vegetarian societies. According to Maurer, vegetarian ideology "provides both a critique of meat-eating and a vision of a vegetarian world" (2). The vegetarian movement's ideology is based on three core tenets that vegetarianism supports: (1) human health, (2) compassion for animals, and (3) environmental sustainability (Maurer 2002, 71). Most vegetarian activists also believe that promoting a gradual dietary transition leads to more permanent vegetarianism for a convert than does encouraging immediate dietary changes. While advocacy organizations tend to agree on the merits of all these tenets, they sometimes disagree on how to present them to the public. For example, their advocacy materials may choose to promote one benefit over others, or they may shy away from the word "vegan" as it is less familiar and may seem extreme to the general public. Some even opt to replace the familiar but culturally-loaded term "vegetarian" with the more benign and factual term "plant-based diet" (Maurer 2002).

Similar to Tom Crompton's (2008) recommendation to privilege altruistic appeals over self-interested appeals, a core framing consideration within vegetarian advocacy is whether to promote altruistic ethical benefits or whether to promote individual, human health benefits. Maurer (2002) found that, for wider appeal, vegetarian campaigns often chose to emphasize health (self-interest). Yet she cautioned that this dietary focus can lead to a loss of integrity over the meaning of the term *vegetarian*, as the public may mistake it as meaning someone who does not eat unhealthy red meat or one who eats very little animal flesh.

Conversely, a campaign that promotes a strong vegetarian identity based on ethical principles, for other animals or the environment, can be inspirational at creating a stronger commitment than a more vague and mainstream appeal to a healthy plant-based diet, but it attracts fewer people (Maurer 2002). Yet, if vegetarianism becomes just another healthy lifestyle choice for consumers, it loses its ideological edge; vegetarianism should be seen as a "public moral good" (126) if it is to become more ideological and foster greater commitment. Maurer explained:

> Many vegetarian leaders seek to move health-motivated, self-interested "exemplary" vegetarians to a more ethical focus that centers on caring more about other humans and animals. This deepening of motivation they see as being key to sparking a greater interest in vegetarian advocacy. (121)

Similarly, Ball and Friedrich (2009) do not think the self-interested appeal (especially to health) is the strongest means to an end because only the anti-cruelty appeal helps to "change the nature of the human relationship with other animals" (41). And in thinking of the end result, they shared that "decades of activism have shown us that the cruelty argument is more likely to galvanize people to action than environmental and health arguments" (42). And no matter what people's excuses are for eating animals, "nothing counters the simple, undeniable fact that eating animals causes unnecessary suffering. We have found it to be the strongest argument for vegetarianism" (45). When health arguments are used, they cautioned for activists to be honest about some of the nutritional needs people should meet to be healthy as vegans, and for activists to avoid exaggeration in general, focusing on credibility.

In keeping with ethical advocacy, activist messages should be authentic to the organization's beliefs. Maurer (2002) mentioned the conflict that vegetarian leaders face when communicating an altruistic animal-focused message internally to dedicated members to retain commitment, while using a more individualistic/health message externally, for utilitarian purposes, to gently attract a wider range of new members. Based on this dilemma, Maurer claimed, "vegetarian advocates must walk a fine line, balancing practicality and moral consistency" (128). She cited activist Jim Mason's description that vegetarian advocacy must avoid seeming fanatical and purist on one end and hypocritical on the other.

Maurer (2002) suggested that the vegetarian movement will not significantly increase the number of vegetarians unless it proves that meat either is dangerous to one's health or is immoral. Since meat consumed in small quantities is not extremely dangerous, it seems to me that the ethical argument is the most compelling option. So, perhaps it is advantageous that a significant portion of the vegetarian movement is comprised of animal rights organizations whose campaigns tend to promote more ethical urgency and inspiration than do the campaigns of vegetarian societies.

Conclusion

As members of a counter-hegemonic social movement, animal rights organizations participate in the struggle over discourse via their strategic use of framing to package their non-speciesist ideologies in ways that resonate with a largely speciesist American public. Activists are faced with the challenge of redefining accepted practices, like meat-eating and animal farming, into socially unacceptable practices. In redefining the status quo as problematic, activists must decide how to balance the risks and benefits involved with either being too critical and oppositional or too moderate. They also must decide whether to base their appeals on altruism and ethics, which match their own altruistic motivations and ethical beliefs, or on human self-interest, which can potentially gain wider acceptance. Lessons from past social movements can be applied to current discussions among animal rights activists over the place of farming welfare reforms and meat-reduction in enacting vegan goals. This shift toward farmed animal welfare reforms has sparked debate within the movement over effectiveness, authenticity, and integrity in social movement strategy – issues central to what this book addresses.

The next part of the book introduces my empirical study of five American animal rights organizations' farmed animal advocacy campaign materials. I outline how they construct various collective action frames that identify problems, propose solutions, and appeal to values that will motivate people to be part of that solution.

Part II:
How U.S. Animal Rights Organizations Frame Food Campaign Messages

Chapter 4

Defining Problems and Culprits, Proposing Solutions

What do animal rights organizations define as the problems with raising and eating animals, and what specific solutions do they propose? To answer this question I analyze the food-related advocacy materials that animal rights organizations create to educate the public about vegetarianism/veganism and the human practices of farming and fishing.[1] This includes those factory farming pamphlets that leafleters hand you at subway stations, the t-shirts and bumper stickers you see vegetarians sporting, the undercover videos of slaughterhouse cruelty that are hard to watch, the vegetarian starter guides and posters displayed at activist tabling events, and "go veg" billboards you drive past along the highway. Other important outreach materials included in my sample are the activists' TV and magazine advertisements as well as their main electronic materials, like websites and self-produced videos.[2]

---

[1] A version of this chapter is published as a journal article in *Society & Animals* (see Freeman 2010).

[2] I found my note-taking on the activist websites was especially laborious due to the wealth of information and many layered pages and links they contained. I admit that spending so much time experiencing hours of video footage and hundreds of photos of animal suffering in factory farms and slaughterhouses took an emotional toll and made me sympathize with the organizations' employees who deal with these issues, images, and animal victims on a daily basis.

To decide whose veg advocacy materials I would study for my in-depth qualitative analysis in 2008,[3] and to compare apples to apples, I selected animal protection organizations in the United States that all have: (1) an animal rights mission supporting veganism (in contrast to a more moderate welfare mission primarily promoting "humane" farming), (2) campaigns providing a variety of food-related advocacy pieces aimed at the public, and (3) a national or international presence in scope. The following five organizations, listed here from largest to smallest, most fully met this criteria: People for the Ethical Treatment of Animals (PETA), Farm Sanctuary, Farm Animal Rights Movement (FARM), Compassion Over Killing (COK), and Vegan Outreach.[4] Before collectively detailing and critically evaluating their strategic communication choices, I'll tell you more about each organization.

---

[3] My method of analyzing and categorizing activists' frames is guided by Stuart Hall's (1975; 1997) cultural studies approach to analyzing texts. This involved systematically examining all words and images in context to uncover their themes of emphasis and their assumptions grounding the social construction of ideas/truths. I often asked, "what is the intended meaning or dominant reading of this message?" and "what does this message assume about the reader and in what position does it situate him/her?" I considered the framing choices not only in context of American cultural values but also in context of their alignment with animal rights ideology.

[4] I recognize that this study cannot claim to represent the whole animal rights or vegan movement in the United States, as it excludes local grassroots groups in favor of a handful of the most professionalized organizations, whom social anarchists might characterize as bureaucratic or even conservative or co-opted (see Torres 2008). Part of this sampling limitation can be excused by my qualitative method that necessitates a smaller sample (depth over breadth). But I also chose these organizations because as an individual grassroots activist in Florida and on college campuses, I found these organizations to be helpful in terms of supporting my campaigns, often by providing free or low-cost materials to distribute. As an activist, I always chose whatever organization's brochure, sticker, video, or poster was most useful to my own campaign strategy and organizational mission, and I often supplemented with home-made materials to have more rhetorical control. I do see these professional organizations as agenda-setters for the movement, not just with the media and government but also with grassroots activists. So I believe they are an important site of study and target of strategic communication advice.

Animal Rights Organizations in My Study Sample

People for the Ethical Treatment of Animals (PETA) (peta.org)

PETA is the largest animal rights group in the world and one of the most well known. PETA was founded by Ingrid Newkirk and Alex Pacheco in 1980 in Washington, DC, after the founders conducted an undercover investigation of a primate research lab resulting in the first ever conviction of an animal researcher for cruelty (PETA History 2012). Now headquartered in Norfolk, Virginia, and run by founder Ingrid Newkirk, PETA has expanded to include international offices and boasts more than 3 million members and supporters (PETA Mission 2012).

PETA overtly identifies with animal rights, stating they are "dedicated to establishing and defending the rights of all animals. PETA operates under the simple principle that animals are not ours to eat, wear, experiment on, or use for entertainment." PETA explains their core stance, saying:

> Animals have rights and deserve to have their best interests taken into consideration, regardless of whether they are useful to humans. Like you, they are capable of suffering and have an interest in leading their own lives. The very heart of all of PETA's actions is the idea that it is the right of all beings—human and nonhuman alike—to be free from harm. (About PETA 2012)

This explains PETA's decision to focus attention on sites where "the largest numbers of animals suffer the most intensely for the longest periods of time: on factory farms, in laboratories, in the clothing trade, and in the entertainment industry" (About PETA 2012).

PETA attracts more media attention than most animal protection groups. In 2011, the media interviewed their staff over 4,500 times and published over 1,000 opinion pieces. PETA had revenues of over $33 million, employed more than 150 full-time paid staff, sent action alerts to more than 1.6 million email subscribers, had more than 40 million people visit their websites, and filled over 650,000 requests for vegetarian starter kits (PETA Annual Review 2012).

Farm Sanctuary (farmsanctuary.org)

Farm Sanctuary is a rescue, education, and advocacy group that "works to protect farm animals from cruelty, inspire change in the way society views and treats farm animals, and promote compassionate vegan living" (Farmsanctuary.org 2012). Farm Sanctuary is unique among my sample organizations because they are a rescue organization, and in fact that is how they got started in 1986 when founders Gene Baur and Lorri Bauston rescued a sheep, Hilda, off of a "deadpile" at an auction and nursed her back to health in their apartment. Their founding purpose is to: "combat the abuses of factory farming and to encourage a new awareness and understanding about 'farm animals.' At Farm Sanctuary, these animals are our friends, not our food" (About Farm Sanctuary 2012).

Now under Gene's leadership, the organization has grown into the largest farmed animal rescue organization in the nation, operating two sanctuaries in California and one in New York as safe havens for thousands of rescued farmed animals. Farm Sanctuary has hundreds of thousands of supporters, with over 600,000 people visiting their websites each year. They currently employ 73 paid staff and have revenues of $6.5 million (Farm Sanctuary Financial 2012).

Farm Sanctuary implies animal rights goals by clarifying their belief that there is no such thing as "humane meat." And while they do support some farmed animal welfare reforms, they see reforms as improvements along the path to the ultimate goal of a vegan world (Farm Sanctuary Positions 2012). For example, they work legislatively to protect "downed" (immobile) animals at slaughterhouses, and in 2002 they partnered with the HSUS on the first ever ballot initiative for farmed animals that outlawed pig gestation crates in Florida. They have since championed successful anti-confinement legislation in Arizona, California, and Ohio, protecting calves and egg-laying hens in addition to pregnant pigs (Farm Sanctuary Victories 2012).

Farm Animal Rights Movement (FARM) (farmusa.org)

Located in Bethesda, Maryland and run by founder and President Alex Hershaft, FARM is the oldest animal rights group dedicated to farmed animal issues, growing out of the Vegetarian Information Service in the late 1970's and changing their name to the Farm Animal Reform

Movement in 1981 (About FARM 2012). Within the last decade, the organization changed the meaning of their FARM acronym so the name now stands for Farm Animal *Rights* Movement to better reflect their strategy of promoting veganism instead of industry reform.

FARM is "working to end the use of animals for food through public education and grassroots activism. We believe in the inherent self-worth of animals, as well as environmental protection and enhanced public health" (About FARM 2012). FARM describes their strategic approach as "pragmatic abolition," encouraging consumers and institutions to reduce meat consumption, even if incrementally, with the ultimate goal being veganism, not free-range animal products.

FARM has organized many annual national animal rights conferences and promotes a variety of ongoing campaigns, such as: The Great American Meatout, World Farm Animals Day, Vegan Earth Day, Gentle Thanksgiving, Meatout Mondays, Green Your Diet, and the Equal Justice Alliance (fighting the federal Animal Enterprise Terrorism Act). They gain exposure through letters to the editor, billboards, demonstrations, distributing several hundred thousand pieces of literature annually, and facilitating pay-per-view sites that offer people one dollar to watch a short factory farming video. FARM has eight paid staff and annual revenues averaging over $400,000 (FARM Financial 2012).

Compassion Over Killing (COK) (cok.net)

COK is a small but influential group located in Washington, D.C. with eight paid, full-time staff and a new office in Los Angeles. Founded in 1995 as a high school club, COK has since expanded to include a nationwide focus (About COK 2012), with annual revenues of over $600,000 (COK Financial 2012) and 50,000 supporters. COK's mission states, "working to end animal abuse since 1995, COK focuses on cruelty to animals in agriculture and promotes vegetarian eating as a way to build a kinder world for all of us, both human and nonhuman" (COK.net 2012).

COK has led some successful campaigns exposing factory farm cruelty and deceptive marketing, particularly in the egg industry, and often works to get mainstream restaurants or food producers to offer more vegan products. Due to its success, its original founders were recruited to the HSUS's farmed animal welfare department, and COK has been under the leadership of Erica Meier since 2005. In 2009, COK's "side of truth" vegetarian television commercial won several awards, a "Do-Gooder Nonprofit Video Award" from the Case Foundation and "Most Effective Vegan Campaign" at the national animal rights conference (hosted by FARM).

Vegan Outreach (veganoutreach.org)

Vegan Outreach is a small group whose name reflects their strategic focus on educating people about why they should go vegan. It is run by founders Matt Ball and Jack Norris (the latter a registered dietician) in Tucson, AZ. The organization began as Animal Liberation Action in 1993 before evolving several years later into Vegan Outreach, based on distributing a brochure of the same name (Vegan Outreach History 2012). Vegan Outreach is "dedicated to reducing the suffering of farmed animals by promoting informed, ethical eating" through the production of full-color booklets that vividly illustrate factory farm cruelty. Their main booklets are titled *Even if You Like Meat, Why Vegan?* and *Compassionate Choices* and are widely distributed by volunteer leafleters and other animal organizations (About Vegan Outreach 2012). Matt told me that as of June 2012, more than 18 million booklets have been distributed worldwide, the vast majority in North America, with translations in twelve languages.

Based on a pragmatic approach to social change, they target young adults through an "Adopt a College" program that has resulted in the distribution of over 7.5 million booklets on more than 1,700 college campuses since 2003 (Vegan Outreach Leafleting 2012). Leafleters receive much positive feedback from people who changed their diet after reading a booklet. Testimonials can be viewed on their homepage, and hundreds of people write weekly to request a "Guide to Cruelty-Free Eating." Vegan Outreach operates with four paid full time staff and posted annual revenues of $985,000 in 2011 (Vegan Outreach Financial 2012).

Descriptive Findings From my Study of Activist Framing

If I was just seeking to determine if these food campaigns were more tactically abolitionist than reformist (see Francione and Garner 2010), then findings lean toward abolition (which is to be expected since my sample is of animal rights groups who promote veganism and didn't include animal welfare groups such as the Humane Society of the United States). But in this study I am more interested in determining what kind of persuasive appeal or rationale is used in the campaign rhetoric and the extent to which that rationale is consistent with animal rights ideology rather than welfare ideology. I'm not just interested in *whether* these activists tell people to go vegan, but *why* – on what moral basis. This values-based rationale can be gleaned from how activist rhetoric defines problems and solutions.

So to structure my analysis of the activists' food advocacy materials, I identify the three framing components that social movements use to promote collective action: (1) diagnosis (defining the problem and possibly attributing blame), (2) prognosis (defining solutions), and (3) motivation (appealing to values to encourage collective action in enacting solutions)[5] (Snow and Benford 1988). In the bulk of this chapter, I outline the first two framing components – how activists diagnose problems and propose solutions – including thick description and examples drawn from their advocacy materials. I'll then end the chapter with my interpretation of the implications of these framing choices for animal rights.

Framing Problems

In studying the farmed animal advocacy materials of PETA, Farm Sanctuary, FARM, COK, and Vegan Outreach in 2008, I find they frame agribusiness as problematic in four primary ways, listed here in order of prominence and frequency, although these categories overlap and are often mutually reinforcing. Raising and killing animals for food:

1) Causes animal suffering due to cruelty,
2) Commodifies animals as objects,
3) Harms humans and the environment, and
4) Kills animals needlessly.

---

[5] The third collective action framing component – motivational values – will be addressed in the next chapter.

In this section, I'll review each of the four major problem categories as constructed by the five groups, describing specific examples from their materials.

1. Animal cruelty and suffering

Farm Sanctuary's "20 Reasons to Go Veg for Life" leaflet ranks the top reason as "'food animals' are not protected from inhumane treatment." This is representative of the most prominent problem frame – the cruelty and suffering of farmed animals – used by all groups. Their materials are full of visual and verbal descriptions of land animals' extreme mental and physical suffering in confinement and the painful transport and slaughtering process. The focus is mainly on cruelty in factory farming but also touches on free-range farming, commercial fishing, and aquaculture. Animal organizations often emphasize the words *cruelty* and *suffering*, such as in Vegan Outreach's web address opposecruelty.org and on the covers of their two most popular booklets: "Even if you like meat, you can end this cruelty" and "Why Vegan? Boycott Cruelty." PETA features sea animal suffering more than any other organization, with a lobster sticker saying "Being boiled hurts!" and a website titled fishinghurts.com. To ensure the public that farmed animals experience pain, animal organizations cite scientific evidence and frequently compare farmed animal feelings to those of cats, dogs, or other popular mammals, sometimes humans. Consider FARM's vegetarian postcard, which states, "Animals raised for food are just as intelligent, lovable, and sensitive as the animals we call *pets*."

All animal rights organizations tend to focus on the *worst* cruelties in factory farming, specifically the extreme intensive confinement of hens in battery cages, female pigs in gestation crates, and calves in veal crates, where the animals can hardly move and the pictures are particularly pitiful, showing bars, excrement, chains, and inflamed raw skin. To specifically expose these three confinement practices, Farm Sanctuary has a "Say No To" leaflet series as well as a video narrated by celebrity vegetarian Mary Tyler Moore titled "Life Behind Bars." Foie gras (enlarged duck or goose liver) is another notoriously cruel practice that is particularly targeted by Farm Sanctuary and PETA with brochures and videos showing the emotionally and physically painful force-feeding by pipes and the resulting wounds and premature death it causes.

To emphasize cruelty, immobility is frequently shown, not only animals confined in small cages, but hens painfully impaled by wire or stuck underneath battery cages, as well as "downed" cows and pigs, too ill to walk, languishing at stockyards or being dragged to slaughter. And almost every factory farming discussion describes the many standard procedures and manipulations performed without anesthesia, including: debeaking, branding, castration, dehorning, toe clipping, ear and tail docking, and teeth clipping. Graphic videos allow viewers to hear and see the animals squealing or crying in pain.

As evidence of the poor living conditions and lack of individual medical care, animal rights organizations cite the high mortality rates on the farm or in transport, showing carcasses rotting in among the living. Videos from Farm Sanctuary and PETA reveal workers beating to death animals who are sick or "runts," particularly in the pork, foie gras, and turkey industries. Commercially-useless newborn male birds in the egg and foie gras industries are shown slowly suffocating in trash bags inside dumpsters. And it is common for any focus on slaughterhouses to assure viewers that some of the animals, particularly birds, are fully conscious when having their throats slit, sometimes experiencing scalding tanks and dismemberment; Vegan Outreach booklets cite a slaughterhouse worker describing how cows often die "piece by piece."

2. Commodification of animals into economic objects

To help explain how and why animals suffer, most animal groups, particularly Farm Sanctuary, have messages critiquing how agriculture treats farmed animals like economic objects instead of sentient beings who are individual subjects. For example, Farm Sanctuary's "Sentient Beings" leaflet states "animals used for food in the United States are commonly treated like unfeeling 'tools of production,' rather than as living, feeling animals." Their farmed animal treatment brochure explains that "when they are no longer profitable, they are literally thrown away," providing examples of how it was deemed legal both to throw "spent hens" into wood chippers and to discard male chicks in the garbage "like manure" (the latter term was used by an egg industry lawyer).

To emphasize the commodity status of farmed animals, Farm Sanctuary's video on downed animals explains that calves may sell for "as little as one dollar but can be left to suffer for days" for that dollar. Their dairy industry video explains how "calf jockeys" round up day old "frail calves, some on the verge of death" to make a "quick buck," and shows men dragging calves by ears or legs and wheeling them off in a wheelbarrow. PETA's *Vegetarian Starter Kit* tells the story of a downed cow left suffering all day at a stockyard because staff veterinarians wouldn't euthanize her, lest they damage the "value of the meat;" she was eventually shot by a butcher and "her body was purchased for $307.50." Farm Sanctuary and Vegan Outreach especially like to quote industry representatives who explain they see farm animals as profit-making machines. A popular pork industry quote advises farmers to "Forget the pig is an animal. Treat him just like a machine in a factory."

To visually express the impersonal business of mass-producing animals, animal rights organizations often show factory farmed animals en masse, especially long shots of warehouses revealing a repetitious sea of similar-looking animals. Other times, closer shots reveal that each pig or calf has an ID number above his/her crate or a numbered tag on his/her back or ear indicating they are nothing more than replaceable units. Farm Sanctuary can juxtapose these industry images of *numbered* animals by showcasing *named* Farm Sanctuary residents like Phoebe or Truffles. Farm Sanctuary has an online profile of a cow, Maxine, who escaped slaughter in Queens, New York, and the text calls the barcode sticker on her back "insulting." Farm Sanctuary removed the barcode to transition Maxine from "food animal to beloved resident."

3. Harmfulness of animal products and farming to people and environment

In addition to focusing on farmed animals, each animal organization's vegetarian guides and webpages contain separate sections on health and environment. Messages from all the organizations except Vegan Outreach[6] prioritize human health as a major benefit of vegetarianism, second only to showing compassion for nonhuman animals. Messages cite the American Dietetic Association, presenting evidence of the health risk of standard meat-based diets, which are associated with increased rates of obesity and major diseases. Animal groups often characterize animal-based diets as *un*healthy. For example, COK's and Farm Sanctuary's veg guides say animal products are the "main source of saturated fat and the only source of cholesterol" for most Americans. Farm Sanctuary's guide also debates the bone-building myth of dairy by saying "studies suggest a connection between osteoporosis and diets that are rich in animal protein" due to calcium being leached out of the bones. And COK emphasizes the unnaturalness of humans drinking another species' milk, especially as adults not in-

[6] In the main booklets they mass distribute, Vegan Outreach emphasizes health the least of all the animal organizations (although they believe veganism is healthy and have separate detailed brochures just on vegan nutrition). In Chapter 7, Matt Ball explains their strategic decision, stating they believe people tend to make major dietary changes due more to disgust over animal mistreatment than a desire to eat healthy.

fants. Both guides also list the antibiotic-resistant bacteria strains that are found in animal products. Farm Sanctuary's brochures warn against "harmful pathogens like Salmonella and E. coli," and declare that Mad Cow Disease and Avian Influenza are "sickening and killing" people. PETA is the only group framing health messages around weight loss and sex appeal. PETA's veg guide has a page on weight loss where a medical doctor states that, "meat-eaters have three times the obesity rate of vegetarians and nine times the obesity rate of vegans." PETA associates meat-based diets not only with being overweight but also with increased rates of impotence.

While individual health risks are highlighted, populist public health issues such as world hunger, farm-worker illness/injury, and rural pollution are sometimes mentioned by PETA, Farm Sanctuary, and FARM. FARM has a "Well-Fed World" campaign dedicated to world hunger policy reform, promoting plant-based diets as a key component to reversing starvation rates as worldwide consumption of unsustainable animal products and factory farming increases. PETA's goveg.com offers links on "World Hunger" and "Factory Farming: Poisoning Communities." The world hunger section explains that much of the world's food, even from developing countries, is used as farmed animal feed for Western diets: "instead of feeding the world's hungry, we take their grains and land to feed our addiction to meat, eggs, and milk." PETA's communities section claims contamination from factory farms is "destroying the heartland" and making people in the surrounding areas sick, as agribusiness is "choosing profits over people." Similarly, Farm Sanctuary's factory farming website's "Economy" link describes how corporate agribusiness pollutes rural communities and fails to bring promised economic benefits.

Of increasing popularity is an appeal to people's concerns for how our food choices affect the environment, especially when it threatens human health. PETA, Farm Sanctuary, and FARM produce print and online pieces dedicated to framing animal agribusiness as environmentally destructive, commonly featuring photos of factory farm pipes spewing manure into cesspools. Farm Sanctuary's "Veg for Life" print pieces all mention environmental degradation, using verbs such as *eroded, ruined, contaminated, compromised, mismanaged,* and *ransacked* and declaring that the number two reason to go vegetarian is because "much of our water and fossil fuel supply is squandered for livestock rearing." Farm Sanctuary has a gray brochure titled

"Factory Farming: Destroying the Environment" emphasizing the pollution of nature and our bodies by showing photos of cesspools, chemical plants, pharmaceuticals, and a fish kill. PETA's *Chop Chop* leaflet visually equates a pork "chop" to trees being "chopped," accompanied with details on meat's association with excessive resource use, damage to oceanic life, pollution, and global warming. FARM has a "Bite Global Warming" campaign built around a 2006 United Nations report listing animal agriculture as an even bigger culprit to greenhouse gas emissions than transportation, a fact increasingly cited by many other animal organizations as climate change becomes a critical global issue (FAO 2006).

4) The killing and needless taking of life for food items

Animal rights groups use this frame the least, as the right to life is overshadowed by a more common emphasis on animal suffering; however, here I provide some examples of when groups, particularly PETA, FARM, and Farm Sanctuary, problematize death specifically. PETA's *Chew on This* DVD lists moral rationales for veganism, including: "because no living creature wants to see her family slaughtered," "because no animal deserves to die for your taste buds," "because they don't want to die," and "because commerce is no excuse for murder." PETA often emphasizes in its calls-to-action how "vegetarians *save* more than 100 lives each year." PETA's teen booklet twice mentions that even animals on free-range farms "all have their lives violently cut short." Bassist Mike D'Antonio is also quoted saying, "Why should somebody have to die if I need a snack?" One page is titled "Bottom Line: Meat is Murder;" this retro slogan of the movement was rarely used by PETA and never used by other animal rights organizations.

FARM has a World Farm Animals Day campaign whose purpose is to "expose, mourn, and memorialize the innocent, feeling animals in factory farms and slaughterhouses." The concept of mourning over the dead is highlighted by FARM's use of death toll statistics presented for each species. The term "death toll" is reminiscent of how nations honor lives lost in any tragedy, like soldiers in war, ensuring that each life counts. The text describes the suffering and death as "pointless," emphasizing meat's needlessness, bolstered by the campaign's slogan "Saving billions – one bite at a time." Necessity was mentioned again in several of FARM's Great American Meatout campaign postcards, saying that each vegetarian "saves up to 2,000 animals" from deaths that are "unnecessary." However, the framing around death, specifically, is often overshadowed by the cruelty and suffering frame.

COK and Vegan Outreach also typically emphasize cruelty, but COK occasionally will use the phrase "saving animals," which implies that we should be saving them from death. And Vegan Outreach's *Try Vegetarian* pamphlet asks people to "Spare the Animal," which more directly implies people are sparing animal *lives* rather than sparing animal suffering, especially as photos of living animals are juxtaposed with their resulting meat product. The necessity of killing animals is overtly challenged once in Vegan Outreach's two most popular booklets by quoting an animal scientist questioning our "right to take the lives of other sentient organisms, particularly when we are not forced to do so by hunger or dietary need, but rather do so for the somewhat frivolous reason that we like the taste of meat." Similarly, COK's video on the broiler (meat) industry says chickens are killed merely to satisfy "our taste for meat, eggs, and dairy," which implies their deaths are for luxury not necessity. COK's "Frequently Asked Questions" (FAQ) section explains that humans do not *need* to eat other animals to survive.

Similarly, Farm Sanctuary's FAQ section states people have the "choice" not to kill, as meat isn't necessary for them like it is for some other animals. They also clarify that hens and cows are killed in the egg and dairy industries too. In their "Veg for Life" leaflet, they rank animal death as the 19th reason (of 20) to go vegetarian, saying "nearly 10 billion farm animals needlessly die every year to fuel the food industry." Farm Sanctuary features a newer advertisement and t-shirt using the phrase "End the slaughter. There are lives on the line" to highlight that killing should cease. Conversely, emphasis on the word "life" implies the opposite of death, as in their "Veg for Life" campaign and in stickers that display the animals declaring they want to live and their life depends on you. Farm Sanctuary campaign materials to protect turkeys, in particular, often say "Save a turkey. Don't eat one." Somewhat similar to FARM's idea of mourning the dead, Farm Sanctuary has a tribute section on farmsanctuary.org that memorializes residents of its sanctuaries who have died (of natural causes) with stories that signify that each individual's life mattered (which ties back with why commodification of animals is a problem).

Blame component: agribusiness first, consumers second

Who is blamed for the problems identified? In framing problems with meat, animal rights organizations identify the most blatant culprit as "factory farms," claiming the "agri*business* industry" perpetuates yet hides its cruelty, killing, pollution, and destruction [my emphasis added in quotes here]. Activist messages only sometimes blame "animal agri*culture*" or "free-range" farms; for example, collateral materials from FARM and Farm Sanctuary specifically ask people to fight or stop "*factory* farming," which is a distinctly different message than "end *animal* farming" would be, as it emphasizes industrial farming rather than all farming, like the small or free-range farm.

To a lesser extent, the government and legal system are critiqued for failing to protect farmed animals; only a few of the animal organizations, particularly Farm Sanctuary, propose legal solutions. Most calls-to-action are for consumers to boycott animal products, with the implication that this dietary change is considered more worthwhile than working with untrustworthy industries and ineffectual government regulatory agencies on welfare reforms.

Animal rights organizations do not usually blame meat-eaters directly and typically use language insinuating that consumers are caring people who are innocently ignorant of factory farm cruelty. Yet even these many positive messages about consumers' compassion suggest, only by default, that newly educated consumers would now be guilty of supporting animal cruelty if they continued to buy animal products.

However, a few messages do more overtly accuse meat-eaters of being responsible parties in the problems of animal cruelty and environmental destruction. For example, A FARM poster shows a cow slaughter illustration and reads, "It's a filthy business. They couldn't do it without you," and PETA's *Chew on This* DVD implicates consumers in the problem by declaring: "it's violence you can stop," "no animal deserves to die for your taste buds," and "it takes a small person to beat a defenseless animal and an even smaller person to eat one." While most environmental frames take a positive approach to asserting the "power" consumers have to save the earth, the most accusing environmental messages for consumers come from PETA, whose online environmental section claims "Meat-eaters are responsible for the production of 100% of this waste. Go vegetarian and you'll be responsible for none of it." And PETA's *Chop Chop* leaflet boldly asserts, "Think you can be a meat-eating environmentalist? Think again!" declaring, "There's no excuse for eating meat."

Framing Solutions

So who should solve the problems identified, and how? The most popular solution that animal rights organizations propose is for consumers to eat fewer or no animal products. Farm Sanctuary also promotes farmed animal welfare reforms by government and PETA promotes some voluntary reforms by industry and retailers. This section outlines these solutions as constructed by the five organizations.

Consumers going vegan

By far the most common solution proposed by all animal rights organizations is for consumers to stop supporting animal agribusiness and to go vegan. While every organization uses the term *vegetarian* more often than *vegan*, veganism is implied through all the recipe and product suggestions that contain no animal products. Farm Sanctuary and FARM favor the term *plant-based* in environmental and scientific messages, presumably as it has less political and social identity connotations. Messages at the end of most print materials make these direct calls-to-action to go vegetarian: "Choose veg foods" (COK), "Kick the meat habit," (FARM), and "Go vegetarian" (PETA & Farm Sanctuary). And consider these vegetarian-promoting URLs used by the groups: Meatout.org, Veganoutreach.org, Goveg.com, Vegforlife.org, Vegkit.org, and Tryveg.com.

Consumers reducing consumption of animal products

Much of the time, organizations, especially COK and PETA, are consistent and clear in their solution for consumers to eliminate animal products, but Farm Sanctuary, Vegan Outreach, and sometimes FARM occasionally suggest less sweeping changes, asking that consumers simply reduce the amount of animal foods eaten. For example, FARM has a "Meatout Mondays" campaign (which, despite the title's focus on flesh, still promotes an egg-free and dairy-free diet too). Outside of Farm Sanctuary's "Veg for Life" campaign materials, a vegetarian solution is not specified in some of its factory farming and stockyard cruelty messages. In some cases, Farm Sanctuary requests only that consumers avoid certain *factory farmed* products, such as: pork from farms using gestation crates, eggs from battery-cage hens, foie gras, and veal. Farm Sanctuary's emphasis on promoting "compassionate" choices sometimes leaves the consumer with the option of determining which food items (even animal products) may qualify as compassionate.

Vegan Outreach's most popular booklet *Even If You Like Meat* suggests that consumers *reduce* consumption of animal products, in particular "eggs and the meat of birds and pigs," as those animals suffer the most. The cover requests that people "cut meat consumption in half," and inside it says "opposing factory farming isn't all or nothing" and consumers should "eat less meat to help prevent farm animal suffering." In this way, they avoid using the word *vegetarian*, suggesting individuals just "do what you can."

No animal rights organization ever proposes that people switch to so-called "humane" animal products or "happy meats." All vegetarian starter guides and websites have small sections dispelling the myth that "free-range" farming is truly free-range (mentioning outdoor access may be limited) or cruelty-free (mentioning that these animals still experience painful mutilations and uncomfortable transport and slaughter).

Industry enacting farmed animal welfare reforms

Some Farm Sanctuary campaigns demand legal animal welfare reform to farming methods, making federal and state governments blameworthy for allowing cruelty. In the video *Life Behind Bars,* spokesperson

Mary Tyler Moore proclaims that crates "should be banned in the U.S. as they are in other countries." This is part of Farm Sanctuary's "Sentient Beings" campaign that also seeks improved legal subject status for farmed animals, following Europe's lead. The website explains that they work on state-wide referenda to institute crating bans as well as federal legislation to protect downed animals at slaughterhouses, requiring euthanasia and forbidding sale of their bodies.

Rather than *government*-based legal reform, PETA tends to pressure *corporations* to voluntarily reform. For example, PETA demands less painful killing practices from Tyson chicken farms and kosher slaughterhouses. And they ran major campaigns ("Kentucky Fried Cruelty" and "Shameway") requesting that the Kentucky Fried Chicken fast-food chain and Safeway groceries mandate higher welfare standards from their egg and meat suppliers. PETA often leverages (temporary) consumer boycotts as a way to pressure the corporations to improve conditions for the animals they purchase.

Summary of Findings

In this chapter I introduced the five U.S. pro-vegan animal protection organizations whose 2008 campaign advocacy materials I analyzed according to how they deployed collective action frames to define problems, identify culpable parties, and propose solutions. My findings identified that these organizations framed problems with meat, farming, and fishing as: (1) the suffering of animals due to cruelty, (2) the commodification of animals as objects, (3) the harmfulness of animal agribusiness and animal products to humans and the environment, and (4) the needless killing and death of animals for food products. The blame component was overwhelmingly aimed at animal agribusiness, particularly factory farms, and, secondarily, consumer demand for animal products. I found that to solve all of these problems, animal rights organizations overwhelmingly relied on consumers to become vegan or reduce their consumption of animal products. In some instances Farm Sanctuary and PETA also promoted agribusiness welfare reforms, whether legal or voluntary, as the solution to the main problems of suffering and commodification. No organization promoted switching to meat, eggs, or dairy from less-intensively-raised animals. Overall, the ultimate goal was to phase animal products out of one's diet for reasons of compassion for animals, concern for health and human welfare, and sustainability.

My Evaluation of Activist Framing in Context of Animal Rights

Up to this point, this chapter has mainly been descriptive. However, in this section I become more prescriptive in terms of offering a critical evaluation of the framing choices made by the five organizations. I also suggest ways that problem and solution frames could more logically align with one another, per Benford's (1987) framing recommendations.

My analysis of these farmed animal/food campaigns reveals that most of their messages did support animal rights goals by promoting veganism and a respect for nonhuman animals' status as subjects not objects. Yet many frames used animal *welfare* ideology to achieve animal rights solutions, conservatively avoiding a direct challenge to the dominant human/animal dualism. To explore this tension, I evaluate the campaigns' problem and solution frames in terms of how they either reinforce or contradict each other, such as:

- The lack of alignment between suggesting veganism as a solution to industry's mistreatment of animals rather than industry's existence.
- The commodification problem frame's emphasis on suffering in *factory* farms rather than the objectification inherent in all animal agriculture;
- The problem frame of killing as a good fit with a vegan solution frame if necessity for survival is emphasized;
- The usefulness of the problem frame of "harm to humans and the environment" if anthropocentric concerns are not made the overall priority; and
- The dilemma of whether or not to make consumer guilt a problem frame.

A Vegan Solution as a Fit with Problematizing Agribusiness's Existence Not its Cruelty

The solution frame suggesting that agribusiness should reform makes sense as a logical solution to decreasing the problem of industry cruelty and commodification of nonhumans, since activists primarily blamed those problems on factory farms. But it does not perfectly align with animal rights ideology, as activist solutions promoting welfare reforms still allow industry to exploit animals but asks that they do so in a way that causes animals less suffering. So when Farm Sanctuary sometimes talks about the need for factory farms to enact "humane" reforms and asks consumers to boycott *factory-farmed* products, without mentioning vegetarianism, it does show alignment between a welfare problem and a welfare solution (as it may imply consumers should buy animal products from less cruel farms). But neither frame is specific to animal rights.

In chapter two I discussed the nuances of the animal rights movement debate over whether or not animal rights organizations should promote agribusiness welfare reforms. Some activists and scholars made a utilitarian argument that welfare reforms are a short-term solution to reducing suffering that works in small steps toward the long-term solution of veganism (Park 2006; Singer 2006; Singer and Friedrich 2006). Even Francione (1996) considered banning some factory farming practices to be aligned with incremental abolition but admitted this tactic is more muddled and problematic than the tactic of promoting veganism. However, I think Farm Sanctuary's campaign to ban foie gras may fit Francione's incremental abolition ideal because it is not a welfare improvement but a ban on an entire product and category of farming. Other scholars and activists made a deontological argument that "improving" an exploitative industry is out of sync with animal rights ideology, adding the utilitarian argument that these reforms undermine vegan objectives by assuaging consumer guilt and possibly helping agribusiness become more profitable by appealing to increasingly conscientious consumers (Dunayer 2006; Francione 1996; L. Hall 2006b; Lama 2006; LaVeck 2006a; LaVeck 2006b; Lyman 2006; Mark 2006; Torres 2006).

The latter ideological and practical concerns over animal rights activists promoting welfare reform solutions fits with my thesis that favors authentically representing ideology. Similar to LaVeck

(2006b), I argue that the solution frame of reforming industry (whether explicitly stated or just implied by problematizing cruelty) muddles and weakens the corresponding solution frame of promoting veganism by suggesting that industry can solve the problem instead of insisting that industry *is* the problem. Additionally, activist organizations' emphasis on the frame of animal cruelty and suffering often highlighted the worst or most abusive aspects of factory farming practices. By doing so, they implicitly made less painful or mundane practices of farming animals, such as captivity and use, seem *less* problematic or even *un*problematic by comparison. Therefore, it is illogical that the direct and main solution to the problem of factory farm cruelty and commodification is to stop eating animals, as that problem implies, instead, that agriculture should switch to less cruel and less intensive practices. To frame veganism as a market-based solution to the cruelty problem specific to factory farms, *implies* that once the industry makes some needed improvements to animal treatment, consumers can then resume their financial support of that industry.

The logical solution to a problem frame of poor animal welfare is for consumers to financially support *less cruel animal farms*. But the main problem-solution relationship that was set up in the majority of animal rights organization frames was to end suffering/cruelty via *veganism*. However, the animal organizations often tried to more logically align the problem of cruelty and commodification with the solution of veganism by explaining that the industry *will not* stop its cruel practices because it is untrustworthy, greedy, and uncaring, and, additionally, it *cannot* stop its cruel practices because its profit motive dictates poor animal welfare in order to remain viable in a global market. When animal rights organizations included this *economic* argument against the reality of welfare reform in their messages, veganism (rather than eating "humane" animal products) became the more logical solution to the problem of animal suffering.

However, a vegan solution to cruelty does not make as much sense if it is proposed along with an industry reform solution, as they inherently contradict each other and require more complex explanations to work together. PETA and Farm Sanctuary did sometimes demand industry improvements (welfare solution) along with asking consumers to go vegetarian (rights solution). This combination of rights and welfare solutions might make more sense if the two were explained to be *unrelated*. To do so, activist messages could clarify that veganism is the most ethical solution to the problem of *using* someone as a resource, and industry reforms are a separate solution aimed at mitigating some of the worst farmed animal suffering endured while society at large is transitioning from an animal-based to a plant-based diet (or because consumers who are apathetic will likely continue to demand animal products). However, this explanation was not clarified in PETA's or FS's messages.

Making a Commodification Problem Frame (of All Animal Farming) Fit a Vegan Solution

Rather than the problem frame of cruelty, the animal commodification problem frame used by animal rights organizations could fit more logically with a vegan solution and animal rights ideology as long as objectification is emphasized in a broad sense more so than just emphasizing its resulting suffering; in that case, more types of animal agriculture can be implicated, not just factory farms. However, in many cases, the activists referenced standard factory farming practices to indicate how the mass production of animals commodifies them and profits take priority over welfare, which could implicitly exclude critiquing a small, more "traditional" form of animal husbandry. I draw this conclusion because even when animal rights organizations argued against free-range farming, they often did so based on the argument that most of these farms were not *truly* "free-range," so that still implies that a *true* free-range farm, albeit rarer these days, would not be objectifying.

These commodification frames become more inclusive of problematizing *all* animal agriculture, not just factory farms, when the activists emphasized the subject status and individuality of each farmed animal, especially when they compared farmed animal individuality to human, dog, or cat individuality. This approach was closer to critiquing *all* use of farmed animals as inherently objectifying, since American society does not allow farming of *subjects*, such as humans, dogs, or cats. Therefore, I conclude that the positive frames that emphasized farmed animal individuality and subject status were more in alignment with animal rights ideology than the negative frames that problematized *factory farming* practices primarily on the basis of being cruel or commodifying.

The Problem Frame of Killing as a Fit For the Solution Frame of Veganism

I think the activists' lesser-used problem frame of "killing and taking of life for human food" is the problem frame that best aligns with animal rights ideology as well as fitting a vegan solution. It relates to the former paragraph's discussion in favor of frames that constructed farmed animals' *subject* status as being equal to dogs, cats, or humans, all of whom are not allowed to be killed for food in the United States. Adding a "necessity" angle could bolster this frame, as it makes logical sense that if people do not *need* to eat animal products to survive, then they cannot morally justify the killing of fellow animal subjects. Some animal rights organizations did occasionally mention this necessity angle, or implied it by emphasizing the healthfulness of a vegan diet. But I think necessity should be emphasized as central to determining when the idea of taking anyone's life becomes immoral and when meat does indeed become murder. This necessity angle is supported by Lee Hall's (2006a) contention that animal rights should not demonize predation overall, as predation is required for carnivorous species to survive.

Problematizing a Meat-Based-Diet's Harm to Humans, the Environment, *and* Wild Animals

Activists also argued against animal products by explaining how meat-production (and consumption) is harmful to people and the environment, which is a form of frame extension (Snow et al. 1986) in relating one cause (animal rights) to other seemingly unrelated causes (public health and environmentalism). While self-interested arguments about human health risks (whether it be from eating animal products or from living in an environment polluted by agribusiness) are a legitimate concern that is astute to mention (HRC 2007), I contend that this frame should not be the main concern emphasized by animal activists, and it usually was not, as it is not as directly related to the animal rights ideology that serves as the organizations' primary motivation. Only FARM generally put human health issues above nonhuman animal issues, and PETA and Farm Sanctuary could only occasionally be accused of doing this in certain communication pieces.

Within this harm frame, *environmental* harm has greater potential than human harm to fit an animal rights ideology, especially if "wild" animals and their habitats are emphasized as deserving protection (LaVeck 2006b). I found that all organizations included wild species, especially ocean life, in their environmental sections in addition to mentioning risks to humans. However, I think there is a missed opportunity to emphasize the inherent value of nonhuman animal life in these frames to more overtly connect the notion of animal rights to protecting wild/free animals (not just domesticated animals) from human exploitation or unhealthy interference. From a deontological (means-oriented) communication standpoint, this allows the moral frames supporting veganism to directly support other animal rights issues, or rights for a wider variety of animals, and not just animals who are farmed and fished.

Problem Frames and Their Relation to the Value of Moral Integrity

One issue that I did not explicitly list as a "problem" frame was that of consumer moral integrity being challenged by the practice of meat-eating, particularly factory-farmed products. However, I do identify moral integrity as a prominent *value* that activists promoted (see chapter 5), and moral integrity was central to the motivational messages that urged consumers to go vegetarian. The fact that animal organizations suggested farming practices were out of sync with the public's general concern for animal welfare *indirectly* makes moral inconsistency a problem for meat-eating consumers and necessitates their involvement in the solution so they can obtain consonance and peace of mind. Attaching the moral integrity value to problem frames against animal agribusiness was one way that animal organizations made the vegan solution seem personally relevant to meat-eaters who might be experiencing guilt. This does not suggest that the use of the moral integrity value fully aligns a *cruelty* problem frame with a vegan solution, as conscientious consumers can still alleviate guilt on the suffering issue by choosing animal products from farms they deem to fit their welfare standards.

If activists framed consumer guilt or moral inconsistency *as the problem*, it would indicate that how we humans *feel* about what we do to other animals is more important than the ethicality of what we *ac-*

tually do to them, perpetuating the idea that humans come first. Therefore, I contend that a guilt problem frame would imply that animal farming and meat consumption are not a problem so long as the consumer has no ethical qualms about supporting it. Therefore, it is more prudent to highlight moral integrity as one of our motivational *value*s rather than making a lack of integrity *the problem*. After all, consumer guilt is the problem from the perspective of animal *agribusiness*, but the exploitation of the animals is the problem from the perspective of *animal rights*, and the two should not be conflated. This is an example of the importance of *values* in framing choices.

CONCLUSION

This chapter entailed my description of how the five animal rights organizations defined problems with eating animals and proposed solutions, including my analysis of ways they could more logically align those problems and solutions with each other and with animal rights ideology. In the final chapter of the book, I build on this analysis to make more specific recommendations that supplement (and in some cases substitute) these collective action framing choices. But first, in chapter five, I elucidate the values highlighted by the five organizations to motivate people to care about the problems and to want to be part of the solution.

Chapter 5

Appealing to Values – Constructing a Caring Vegan Identity

Having just explored the problem and solution components of collective action framing in the previous chapter, here I explore the motivation/identity component, manifested in terms of the values that activists construct around veganism and what kind of character they propose that a vegan embodies. In this chapter I will detail sixteen separate values-based appeals that I found the five animal organizations used in their persuasive messages.[1] While I have elected to organize the values in this way, you will see they are not, in actuality, isolated or discrete categories. They are the overlapping and mutually-reinforcing values that I felt were implicitly, and sometimes explicitly, alluded to as being motivational based on how activists chose to frame problems and solutions (ex: by problematizing cruelty, the activists indicate people identify as compassionate). So in my analysis, I asked, "what do the proposed problem and solution frames imply about the kind of values a viewer would need to care about and identify with to be persuaded by this message?" Framing's reliance on activating values reflects the theory that frames are recruiting tools for ideology (Oliver and Johnston 2005). As such, the activists' collective action frames in chapters four and five attempt to work symbiotically, like an ecology, to produce a farmed animal rights discourse.

---

[1] You can also find a version of this chapter published in Emily Plec's edited book *Perspectives on Human-Animal Communication: Internatural Communication*, published by Routledge in 2013. My chapter is entitled "Stepping Up to the Veggie Plate: Framing Veganism as Living Your Values." (see Freeman 2013)

My findings suggest that, to appeal to the greatest number of people, activists attempted to align veganism with identity traits and values that were relatively common and non-controversial in American culture. In this way, veganism is constructed as a natural fit for most people rather than a liberal subculture or extremist identity. According to the five animal rights organizations, the choice to eschew animal-meat, eggs, and dairy potentially says a lot about someone's character. Most predominantly, it says they feel compassion toward fellow animals because they respect animals' status as feeling individuals, and thus want their own actions to be consistent with these values, demonstrating personal integrity. It also says they likely prioritize other altruistic, caring values such as: life, freedom, environmentalism, humanitarianism, and the desire to make a difference. It says they likely believe in ideals such as: honesty, naturalness and purity, patriotism, and populist notions of fairness. Finally, it says they identify with common-sense personal values favoring individual well-being, such as: health, choice, desire for pleasurable and convenient food, and a sense of belonging and attractiveness to others.

At the end of the chapter I evaluate which of these 16 values were most informed by animal rights ideology, or how they could be reframed to better align with animal rights ideology, particularly the values of: compassion, respect for sentience and individuality, moral integrity and consistency, honesty, life, freedom, naturalness, belonging, and desire to make a difference. First I begin this chapter by describing the 16 values-based appeals of activist campaigns, sharing a few examples of how animal rights groups associated each value with veganism. I organize the values into three major categories: (1) altruistic, (2) idealistic, and (3) personal.

Appeals Made to Altruistic Values

In this section I discuss how the five organizations frequently appealed to people's sense of altruism and concern for others (human and nonhuman), including the following values: compassion, respect for sentient beings, life, freedom, humanitarianism, environmentalism, and the desire to make a difference. In this way, they construct veganism as the diet most befitting a responsible, caring citizen.

Compassion and Caring for Nonhuman Animal Suffering and Aversion to Cruelty

All five organizations spend much of their time educating consumers about factory farm and slaughterhouse cruelty, implying consumers are compassionate and caring enough to be offended, an assumption which is also explicitly declared. For example, Farm Sanctuary's new slogan is "a compassionate world starts with you." And PETA's *Meet Your Meat* video ends with celebrity vegetarian Alec Baldwin telling viewers to think about the cruelty they have seen, to choose "compassion," and to go vegetarian as "millions of compassionate people" have decided to do. Vegan Outreach's *Why Vegan?* booklet declares "we can choose to act with compassion by boycotting animal agriculture. Making humane choices is the ultimate affirmation of our humanity." COK's *Vegetarian Starter Guide* creates a good versus bad dichotomy under the title "Choosing Compassion Over Killing" by asking "Do we want to support kindness and mercy, or do we want to support cruelty and misery?" requesting that readers "take a stand for compassionate living." FARM literature often describes people as "caring," such as if someone saves a turkey at Thanksgiving, and FARM's Great American Meatout postcard includes the value of care in the headline, "Because you care about animals."

Respect for the Sentience and Individuality of Other Animal Subjects

Built into the assumption that someone is compassionate toward nonhuman animals is the idea that the person respects the other animals' ability to feel and does not want them to suffer. The concept of sentience, as I use it here, involves not only experiencing pain, but also experiencing emotions, thought, or consciousness. Activists typically do not use the unfamiliar term "sentience," rather they tend to simply say that animals "feel" or "suffer." To emphasize nonhuman animals' status as feeling subjects not commodified objects (someone not something), all organizations use gendered or personal pronouns like *he, she,* or *who* when referring to individual farmed animals. This is unlike the common American practice of calling them *it* – a pronoun denoting inanimate objects.

All five organizations include frequent messages to ensure the public that farmed animals are sentient, often comparing their capabilities to those of cats and dogs or sometimes to other animals, including, less frequently, humans. An example of a pet comparison is FARM's vegetarian postcard, which states, "animals raised for food are just as intelligent, lovable, and sensitive as the animals we call *pets*." An example of a human comparison is PETA's teen vegetarian booklet, which declares, "animals are like *us*" and proceeds to describe farmed animals doing what would normally seem like human activities, such as pigs playing video games, turkeys playing ball, cows babysitting, and fish gardening.

Farm Sanctuary's "Sentient Beings" campaign seeks elevated legal status for U.S. farmed animals, to be classified as sentient beings as they are in Europe. The leaflet for the campaign is titled "Farm animals have feelings too" and says these animals are "sentient beings – capable of awareness, feeling, and suffering" who "deserve to be treated with respect." This is contrasted with pictures of farmed animals in extreme confinement and quotes from industry that compare them to machines and manure, a tactic used by many organizations to demonstrate how industry commodifies animals.

To showcase farmed animals as individuals, all groups portray pictures of animals making direct eye contact with the reader. Every organization uses photos of farmed animals getting their faces close up to the camera as if to indicate their curiosity and sociability. Several vegetarian starter guides describe the personalities of each rescued animal and display his/her portrait and individual name, such as Travolta the cow, Emery the chicken, and Ashley the turkey. The descriptions reveal personality traits, such as friendliness, talkativeness, playfulness, loyalty, and preferences for certain foods such as apples or green grapes. In Farm Sanctuary's *Guide to Veg Living*, a photo shows a goose, Bing, happily spreading his wings in a pond and honking with gusto, and another photo shows a piglet, Rudy, standing proudly and defiantly in the grass with the low camera angle putting the viewer in the position of looking up at him so that he appears larger than life.

Several PETA publications feature an "Amazing Animals" section praising animal abilities. Here are examples of PETA's opening sentence descriptions for each species:

> Chickens are inquisitive, interesting animals who are thought to be as intelligent as cats, dogs, and even some primates; Pigs are curious and insightful animals thought to have intelligence beyond that of an average 3-year-old human child; Fish are smart, sensitive animals with their own unique personalities; Cows are intelligent, loyal animals who enjoy solving problems; Turkeys are social, playful birds who enjoy the company of others; Geese are very loyal to their families and very protective of their partners and offspring.

PETA dedicates more space to fish than does any other group, and they are the only group that talks about fish sentience in terms of intelligence and personality. PETA has a fishinghurts.com website, brochures, and collateral materials dedicated to sea animals. The other organizations who mention fish, particularly COK, Farm Sanctuary, and Vegan Outreach, to a minor degree, often talk about them only in terms of an environmental issue. COK does talk about fish sentience in terms of their ability to feel pain but not in terms of personality.

Some messages overtly request that consumers view farmed animals as more than food objects. The very title of PETA's popular video, *Meet Your Meat,* juxtaposes the idea that consumers can see farmed animals both as individual subjects while alive and as objects after death. COK has a print advertisement displaying a cow's face reflected in a woman's eye and asks teen girls to "see her as more than a meal." Farm Sanctuary's print ad features a young pig, Truffles, who challenges the viewers to "look me in the eyes and tell me I'm tasty," and a sticker showing an illustration of a chicken stating, "I am not your breakfast, lunch or dinner." Similarly, PETA has several collateral materials with an illustration of a chick declaring "I am not a nugget" and telling viewers that pigs and fish are "friends not food." When it comes to pigs, PETA reminds viewers that they have an assumed attachment to one pig as an individual subject – Babe, from the movie of the same name – using a poster that shows a piglet and reads, "Please don't eat Babe for breakfast." Emphasizing friendship, as several organizations do, challenges the solely instrumental relationship we typically have with farmed animals. FARM has a button and t-shirt that declares "I don't eat my friends" and shows an illustration of a man surrounded by farmed animals and a cat. According to Matt Ball, Vegan Outreach likes to use photos of people posing with farmed animals in ways that would seem reminiscent of how people affectionately interact with their companion animals.

Caring for Fellow Human Beings

While animal rights organizations naturally emphasize care and respect for nonhuman animals, some additionally show how veganism is also compassionate towards humans. Consider FARM's t-shirt that says, "Stop human and animal suffering. Go vegan!" This "anthropocentric altruism," as I am calling it, is particularly emphasized by FARM, PETA, and Farm Sanctuary who all have anti-factory-farming campaigns that either fight human hunger, worker exploitation, or the health contamination of rural neighborhoods by pollution (see problem #3 in chapter four). There is an indication that the public values protecting innocent, vulnerable people from being taken advantage of by greedy, powerful corporations. For example, PETA warns "profits are put before people" by government and factory farmers, so readers are encouraged to go vegan to "stop these exploitative industries and promote a world of compassion." These types of problem frames imply that the public values human life, well-being, and an egalitarian sense of justice.

Caring for the Earth and its Inhabitants

Of increasing popularity is an appeal to people's concerns that our food choices work in harmony with nature (and avoid harming its human and nonhuman inhabitants). In chapter four, you learned about how PETA, Farm Sanctuary, and FARM have print and online pieces specifically dedicated to framing animal agribusiness as environmentally destructive and wasteful, commonly featuring photos of factory farming pipes spewing manure into cesspools. In this way veganism is associated with purity, naturalness, and health and wellbeing, while factory farming is associated with the opposite – contamination and pollution. Another key aspect of environmentalism is sustainability, so a plant-based diet is framed as using natural resources wisely and efficiently, while, conversely, meat production is associated with being wasteful and destructive. Many values are embedded in environmentalism, but I think the most overarching value is a sense of responsible use of resources.

Life

To show a value for the life of all living beings, animal organizations sometimes specifically problematize the killing and death of nonhuman animals, not just their suffering. To promote saving life, COK names its main veg starter guide sections: "Saving Ourselves," "Saving Animals," and "Saving the Earth." And consider the use of the word *life* in FARM's "choose life" slogan for its Great American Meatout and in its thanksgiving campaign declaring that killing innocent animals "betrays the life-affirming spirit" of the holiday, asking viewers to "celebrate life." Farm Sanctuary sells a t-shirt that has a quote by Buddha which reads, "All beings tremble before violence. All fear death. All love life." In a direct life-saving appeal, Farm Sanctuary has stickers showing cows and chickens with a statement reading, "She wants to live and her life depends on YOU!" Animal groups often talk about the number of animal lives saved by vegetarians per year, or conversely, how many lives meat-eaters take.

Veganism is also framed as saving or extending the lives of humans. In Farm Sanctuary's "Veg for Life" campaign, one way to interpret the meaning of the word *life* is that a healthy vegetarian diet can save one's own life and/or the lives of farmed animals. And the main vegetarian guides for Farm Sanctuary and PETA title the recipe section "Recipes for Life" to indicate food choices that result in healthy, living bodies. Life could also connote time, suggesting people should eat vegetarian for the rest of their (long) lives.

Freedom

Freedom is a key American value and a popular part of political rhetoric, so animal organizations attempt to apply it to *all* animals having the right to freedom over their own body and space. PETA's *Chew on This* DVD declares "everyone wants to be free," meaning nonhuman animals too. But, besides this example, *freedom* as a term is more often implied than it is directly referenced. Activists' consistent emphasis on animals' extreme confinement and immobility in factory farms suggests that Americans will find these high levels of restriction to be unfair.

Although infrequent, direct analogies to human confinement are sometimes made, such as COK's pork leaflet comparing the confinement of pregnant pigs "stuck" in gestation crates to how frustrated and uncomfortable people would feel being "stuck" in a car in traffic for years. It says pigs are "unable to move freely" and "can't even walk or turn around," as they are in a pregnancy cycle going between gestation and farrowing crates. The word *cycle* is used by many organizations to describe the treadmill of re-impregnation faced by pigs and cows to maximize industry profit.

Freedom is associated with wide-open spaces, as in the American West, which may be why Farm Sanctuary emphasizes space on its new home page, with lots of white space, a picture of rescued animals enjoying the sun, and a strip of grass across the bottom. Blue sky, sun, and grass are often represented in all animal organizations' pictures of contented animals to emphasize their relative freedom (presumably in a sanctuary) in contrast to the darkness, filth, and discomfort of captivity on industrial farms.

Besides wild-caught fish, it would be complicated for animal rights organizations to claim that animals commonly used for food should be set "free," as animals raised in captivity are domesticated and most may not be able to survive in the wild. But Farm Sanctuary sometimes uses the word "free" when describing how rescued animals in the sanctuary are free from pain and free to roam outdoors and enjoy life. Yet, farmed animals on the few small farms that are truly "free-range" might have similar space to those at Farm Sanctuary, but photos of these few "free-range" farm animals are not used in animal rights literature, as they would likely fail to promote the same level of contempt for captivity and agriculture as do the pictures showing animals in intense confinement on factory farms. However, viewers of activist materials may not know for certain that the contented animals shown in pictures are not just from "better" farms, which may lead viewers to immediately think of animal freedom primarily in terms of "free-range" farming not freedom from human control.

Desire to Improve the World and Make a Difference

Activist messages indicate that people seek meaning in their lives by improving the world and making a positive impact. COK encourages readers to "Make a difference. Start today!" and the back of its vegetarian starter guide states in bold "every time we sit down to eat, we can make the world a better place." Similarly, Vegan Outreach's *Even if You Like Meat* booklet tells readers "every time you choose compassion, you're making a difference." To emphasize personal empowerment, FARM uses the slogan, "Stop global warming one bite at a time," describing the "power" of our food choices and how they "matter." Similarly, Farm Sanctuary has a radio PSA for Earth Day that says the "power is on your plate" to protect the earth "every time we eat." Note that in many of these slogans there is a time element emphasizing the ease with which a person can make a difference through vegetarianism *everyday* because it allows him or her to improve the world "at every meal" or "one bite at a time." Eating is a mundane and convenient form of activism for those who do not necessarily want to dedicate time to being a traditional activist or do not have much money to donate to causes.

Further indicating the importance of a switch to vegetarianism, animal organizations often claim that vegetarianism is the *best* way to help animals and the planet. Farm Sanctuary's *Veg for Life* brochure says, "eliminating ALL animal foods from our diets is the single most important step we can take to be kinder to animals, ourselves and the Earth." And PETA often cites vegetarian rock icon Sir Paul McCartney telling readers, "If anyone wants to save the planet, all they have to do is just stop eating meat. That's the single most important thing you can do."[2] Vegan Outreach's *Why Vegan?* booklet also emphasizes impact by saying, "over the course of a lifetime, one person's food choices affect hundreds of animals."

Making a difference is also connected with self-interest in feeling good about oneself. PETA's online "pledge to be veg" appeals to people who want to make things "better" and do the right thing, by having them agree, "I want to eat better, feel better and stop supporting cruelty." This could imply feeling better both physically and mentally. Similarly Farm Sanctuary predicts that through veganism you will "feel good because you make the world better," stating that vegetarians enjoy better "mental health and feel good knowing they are working toward improved health and well-being for themselves, animals and the environment." Regarding mental health, veganism is often framed as a personal growth goal. COK's veg guide section on transitioning to a vegetarian diet reads like a life coach's plan praising new

---

[2] Although, I think it's important to note that many behavioral and institutional changes are necessary to "save the planet," even while acknowledging the ecological importance of transitioning humanity to a plant-based diet.

vegetarians because they have "made it!" and deserve a "pat on the back!" Farm Sanctuary's veg guide has a section on transitioning one's diet that indicates that vegetarianism is worth the effort by telling readers to, "give yourself a break" if every move is not perfect, encouraging readers that "every step you take to reduce suffering, exploitation and injustice is always a step in the right direction."

APPEALS MADE TO IDEALISTIC VALUES

While similar to altruistic goals, this next set of values I am characterizing as "ideals" as they represent socially-respected goals or virtues, including: honesty, naturalness, populist notions of fairness, and patriotism. In this way, animal organizations broadly associate veganism with positive moral virtues of truth, loyalty, fairness, and purity.

Honesty

Consumers and citizens need truthful information to make informed decisions, and activists blame agribusiness for misleading consumers and hiding the ugly truth of factory farming; animal organizations see it as their job to give people a reality check. COK emphasizes the honesty aspect the most with their campaign against fraudulent "humane" farming labels, particularly exposing the egg industry's "animal care certified" logo. And they also feature television ads showing flabbergasted consumers at a fast food restaurant being served a rare "side of truth," hearing the miserable conditions faced by the animals who make up their meal.

In a unique twist on honesty, PETA's *Chew on This* DVD accuses *parents* of being dishonest to children when the narrator says, "you shouldn't have to lie to your kids" about where their food comes from. This assumes that adults know that the reality of farm animal suffering and death is gruesome enough to upset kids and possibly keep them from eating meat. To some extent this tactic belies the common activist frame that meat-eaters are simply innocent, caring people uninformed about the mistreatment of farmed animals.

One could even interpret the animal rights organizations' frequent use of referenced citations and photos as a way to emphasize that they themselves are telling readers the truth about the poor conditions of factory farming and the environmental and health risks associated with animal products. Since it is clear that these advocacy materials are partisan, activists often cite outside experts, especially with doctorates, to prove the truth of statements in favor of vegetarianism or agricultural practices. Vegan Outreach is the most meticulous in its use of other experts and inclusion of detailed citations within the document to add validity. All the vegetarian starter guides cite outside scientific sources, particularly in the sections on environment and human health, and also to prove claims of the sentience or capabilities of other animals.

Naturalness

Naturalness is a value the public relates to food's healthfulness, in contrast to artificial foods. For example, PETA and COK frame cow's milk as unnatural for humans (and all adult mammals) to consume. And COK's *Vegetarian Starter Guide* contains a section titled, "Is milk natural?" and shows an illustration of human children sucking on a cow's udder, the absurdity of which indicates the answer is *no*. It explains how each species' milk is "intended" for its own species and that humans are the only species who drinks the milk of other species. Similarly, PETA's veg starter kit also says, "no species naturally drinks milk beyond the age of weaning" and PETA's *Meet Your Meat* video emphasizes how the cow naturally gives milk for her baby not for humans. In contrast to dairy, *meat* was only once (by PETA) framed as unnatural for humans to eat. In the health section of goveg.com, "Is eating meat natural?", PETA argues that, among other physiological reasons, humans are not as naturally equipped as carnivorous animals to kill and eat raw animal flesh with their bare hands.

In addition to the health angle of naturalness, animal organizations place greater emphasis on what is natural for chickens, pigs, and cows as opposed to what is artificial or industrialized, such as the genetic modification of animals and large-scale factory-farming. For example, animal organizations often show photos of farmed animals in a more natural state outdoors in the sun, contrasted with images of them behind bars, often in the dark of a warehouse, looking unkempt and crowded on factory farms. Mechanization, metal, and concrete abound in images of animals confined in warehouses, left to die in garbage cans, and disassembled on the slaughterhouse assembly line. The feel is cold, dark, gray, dirty, and industrial. This unnatural environment is juxtaposed against the cleanliness and brightness of portraits of pigs, cows, and birds, presumably in a sanctuary or backyard farm, surrounded by natural elements of sun, grass, hay, wooden fences, and ponds. To a lesser degree, wild animal species are shown living in nature, particularly fish and sometimes turkeys.

To label factory farm practices as unnatural or untraditional, Vegan Outlet booklets use a headline that reads "not your childhood image," and both PETA and COK use messages aimed at youth that state, this is not "Old MacDonald's" farm. These tactics indicate that viewers expect or want farmed animals to live a life outdoors or in a clean barn – a life that is closer to what would be considered "natural" for them, either natural for their species in the wild or natural for old-fashioned animal husbandry. Animal organizations often directly refer to practices, conditions, and the animals' bodies being unnatural in modern animal agribusiness. For example, Farm Sanctuary says the number 10 reason to go vegetarian is because "farm animals are usually prevented from engaging in instinctual behavior and live a fraction of their natural lives." And Vegan Outreach's booklets cite Michael Pollan saying of a battery-caged hen that, "every natural instinct of this animal is thwarted." These stifling factory conditions are contrasted with the descriptions of how these species would behave in nature.

Another highlighted aspect of unnaturalness is the unnaturally large weight the industry demands of animals grown for profit and how artificial and painful this immense girth can be for the animals. Farm Sanctuary's video on the turkey industry explains how farmers alter the shape of the birds to meet consumer demands for turkey breasts, stating that this "anatomical manipulation" has made male turkeys so large that it is impossible for them to "mount and reproduce naturally," so females must be "artificially inseminated" by human farm workers. And COK's veg guide explains how birds "grow so abnormally fast due to selective breeding and growth-promoting antibiotics" that they suffer organ failure and lameness, sometimes collapsing under their own bulk.

Many organizations critique the unnatural diet and medications that agribusiness uses to fatten animals, likely spurring our own concerns over the health threats of frankenfoods and superbugs. Chickens are "dosed with a steady stream of drugs" (PETA) and cows are "fattened on an unnatural diet of grains and 'fillers' (including sawdust and chicken manure)" (COK). Farm Sanctuary blames bovine growth hormone injections for why cows "produce ten times more milk than they would in nature," showing engorged udders practically dragging on the ground. They also mention here that dairy cows are killed at the young age of four or five when they could live to be 20 years or more. Sometimes PETA also highlights the animals' premature death, which could be considered an emphasis on how animals do not get to live out their "natural" lifespan when farmed.

Similarly, all the environmental messages about how animal agribusiness heavily pollutes the land, air, and water seem to be based on the belief that nature is clean but can be contaminated and become toxic or ruined by humans. These environmental messages from animal organizations highlight how such out-of-sync agribusiness practices contaminate the purity of nature. Farm Sanctuary's factory farming brochure cites the Worldwatch Institute saying, "overgrown and resource intensive, animal agriculture is out of alignment with the Earth's ecosystems." To further indicate artificiality, the brochure shows photos of pharmaceuticals as well as fumes emanating from an agricultural chemical plant. Related to this, the vegetarian guides of Farm Sanctuary, COK, and PETA all mention contamination in the resulting animal products humans eat, saying how animal products are a health risk because they contain unnatural ingredients like pesticides, drugs, and other chemicals.

Fairness: American Populism and the Responsibilities of Big Business and Government

This broad category of populist notions of fairness overlaps with the values of naturalness, honesty, and concern for human wellbeing, as activist messages capitalize on an assumed public mistrust in the exploitative and irresponsible tendencies of big business and, in some cases, government. This idea of American populism suggests that animal organizations assume people want corporations and the political elite to be held accountable in cases where they take advantage of the innocent little guy. For example, all animal organizations critique modern agricultural practices specifically on the basis that it is contemptible as "factory farming," "corporate agribusiness," or an "exploitative industry." This contrasts with the bucolic values that consumers may have for wholesome traditional or family farming, considered a responsible business of everyday hard-working people. Hence, activists rhetorically focus blame on agri*business* more than agri*culture*. Activists generally do not insinuate that small or "family farms" are nearly as problematic. Factory farming, in particular, is largely to blame for why cruelty is standard, food is unwholesome, the earth is polluted, workers are exploited, and consumers are misled.

PETA and Farm Sanctuary appeal to these populist values the most, as they both have online sections discussing the exploitation of workers and the contamination of rural communities by animal agribusiness. The implication is that industry is greedy and callous and fails to demonstrate justice, respect, responsibility, and decency toward the common man. The jobs agribusiness provides are described as dangerous, dirty, and low paying. PETA cites workers who explain how their bosses cheat them out of wages and workers' compensation, firing those who complain. To highlight objectification, PETA quotes a farm worker saying he felt he was "disposable" and treated like a "machine," and a contract chicken farmer said she was "treated like a dog" by the industry. To further emphasize worker mistreatment, PETA shows pictures of working class people protesting and striking and describes industry as anti-union. In this section, PETA also occasionally uses trigger words for exploitation like *serfs, slaves,* and *child labor.* This is contrasted with wholesome "community" values of rural America, or the "heartland," where people simply expect basic, fair treatment from employers and a safe, healthy environment for their families and community. Farm Sanctuary's section on the economic issues of factory farming laments the loss of family farms, saying, "small farms help to create close-knit communities and thriving local economies."

Perhaps surprising for such a liberal animal rights organization, PETA's sections on the polluting of rural communities and the negligence of government might also appeal to politically conservative values, especially those that mistrust the federal government. Because most animal organizations propose a consumer solution instead of a government solution, this could be construed as valuing the notion of personal responsibility, consumer choice, and free market capitalism. For example, PETA's page on government negligence shows a photo of the Capitol Building in Washington, D.C., specifically emphasizing *federal* government agencies, such as the United States Department of Agriculture and the Environmental Protection Agency, and not implicating *local* governments. Government regulation is portrayed as futile because industry lobbying and money has corrupted the process, so consumers, more so than government, must right the wrong by boycotting factory farming.

Patriotism and American Pride

Although it isn't a common theme, American pride is sometimes directly referenced through the use of patriotic symbols, such as PETA's bumper sticker displaying an American flag and declaring, "Proud to be a Vegetarian American." And FARM's Great American Meatout campaign uses red, white, and blue colors as found in the American flag. One of its posters has Uncle Sam, portrayed by a cow, pointing at the audience, reminiscent of the iconic war recruitment poster, with this one saying "I want you to stop eating animals." The text emphasizes loyalty by stating viewers should join the meatout "for your honor, for your family, for your country, and for your planet." Perhaps some animal organizations feel that noting the patriotism of veganism helps make it more mainstream to American culture and less "extreme" or radical, something particularly useful in a political era where being a radical or extremist is akin to terrorism (see Potter 2011).

Sometimes the animal organizations give an indirect nod to American pride by suggesting the farmed animal protection policies of the U.S. government lag behind those of other countries, perhaps invoking a sense of team spirit to improve America's reputation and become more morally progressive, befitting of America's self-imposed identity as the greatest nation on earth. This strategy of comparing animal protection laws internationally is used most frequently by Farm Sanctuary because they have some of the only campaigns calling for federal legal reform of industry. Farm Sanctuary's *Eggribusiness* video explains that European nations have already outlawed battery cages, so "It's time for birds to be protected from abuse in America too." And in Farm Sanctuary's *Life Behind Bars* video, celebrity spokesperson Mary Tyler Moore informs viewers that legal protection for American farmed animals is "grossly inadequate." She states gestation crates, battery cages, and veal crates should be banned in the U.S. as they have been in Europe. The call-to-action is that America has an "ethical obligation" to prevent animal suffering as a "civilized nation."

The video also cites U.S. Senator Byrd critiquing factory farming as "barbaric" and saying a "civilized nation" must be more "humane" toward life.

Related to America's notion of itself as a civilized society is FARM's use of caveman analogies in two cases, implying that if people are still eating or wearing animals in the 21st century, they are uncivilized and undeveloped. While this could suggest that people simply have more options in modern times, it also capitalizes on Americans' views of themselves as citizens of one of the most highly-developed, advanced, and industrialized nations. Perhaps ironically, it could suggest that Americans no longer behave like the *animals* that were their primitive ancestors. Similarly, Farm Sanctuary appeals to America's pride with a t-shirt bearing Gandhi's quote, "The greatness of a nation and its moral progress can be judged by the way its animals are treated."

APPEALS MADE TO PERSONAL WELLBEING VALUES

In addition to appealing to altruistic and idealistic values, activists appeal to common-sense *personal or individual values* that emphasize the self-interested benefits that a person receives to his/her own wellbeing by going vegan. These values include a desire to be healthy, have lots of choices, be attractive and socially accepted, enjoy delicious food that's convenient to access, and feel a sense of pride in one's moral integrity. In this way, veganism is constructed as the best dietary choice for someone's physical and mental wellbeing.

Human Health

Second only to showing compassion for nonhuman animals, activists prioritize health as a major benefit of vegetarianism. This value is practical in the sense that, at its core, food is necessary for survival and people need to be assured that a plant-based diet contains the basic nutrients needed for humans to thrive and live a long healthy life, as this may run counter to mainstream beliefs in the old "four food groups" nutritional model from the U.S. Department of Agriculture (USDA) that used to recommend dairy and meat as necessary categories for the human diet.[3] Vegetarian messages must then address and overcome these common nutritional myths regarding the perceived necessity of animal protein.

Eating a plant-based diet is shown to reduce one's risk of getting sick, as animal organizations focus a lot on meat and dairy's link with major diseases like diabetes, heart disease, stroke, cancer, and osteoporosis. So a plant-based diet is positioned not just as nutritionally adequate but as healthier than the standard American animal-based diet. They often cite the American Dietetic Association's positive position on vegetarian diets emphasizing the disease-fighting properties of plants, as they contain fiber and antioxidant nutrients as well as being leaner and artery-friendly.

---

[3] Recently the USDA has moved from a pyramid to a plate framework for modeling nutritional guidelines and recommends the plate be half filled with fruits and vegetables, with the other half filled with whole grains and lean proteins (USDA 2012). Notice that the protein category is not labeled "meat," as it includes plant-based proteins such as beans, peas, nuts, and seeds, including soy products, in addition to animal flesh and eggs. And the protein section includes a vegetarian tipsheet acknowledging non-animal proteins are healthful, stating "Protein needs can easily be met by eating a variety of plant-based foods" (USDA Vegetarian 2012).

There is an animal-based food category labeled "dairy" that the USDA model shows as a beverage next to the plate. I suggest it could be re-labeled as the "calcium" category, as that is the main focus of the nutritional guidelines there, and it includes calcium-fortified soy beverages (non-dairy). The recommendations lean more toward cow's milk and they seem to suggest that those who don't eat dairy do so because they are lactose-intolerant/allergic, never mentioning ethical reasons. The dairy section provides suggestions for how to get around the lactose allergy rather than promoting non-dairy alternatives. The section seems to also steer readers away from other non-dairy calcium sources as being somehow deficient, stating "calcium-fortified foods and beverages such as cereals, orange juice, or rice or almond beverages may provide calcium, but may not provide the other nutrients found in dairy products" (USDA Dairy 2012). The USDA does not mention the reverse logic that dairy products such as cheese probably don't contain all the other important nutrients found in orange juice, cereal, or nutmilks. And despite the high calcium content of leafy greens, one has to search a bit further among the dairy links to find out that they contain calcium. Overall, it is unclear why dairy or even calcium merits its own food category, as no other food category is designed around one nutrient. I predict that in years to come, despite the dairy lobby, the USDA will not be able to continue to justify why dairy is designated special treatment in the U.S. nutritional guidelines, as 1) many people are allergic, 2) drinking milk (especially of another species) as adults is unnecessary and unnatural among mammals, and 3) calcium can be found in more absorbable sources, especially leafy greens.

Health as a value overlaps with other values of *naturalness* and *desire for belonging and social appeal*. For example, plant-based foods seem more natural, fresher and safer than animal-based foods that contain antibiotics and other pharmaceuticals, such as hormones, as well as gross contaminants such as pus, bacteria, and pathogens that cause potentially fatal food-borne illnesses. In this way, they seem purer and more wholesome. And a healthy diet is also connected with one's motivation to be sexually active, fit, and attractive to others. PETA in particular emphasizes that vegetarian diets help one to be thinner and more energetic, which is seen as sexier than being overweight, sluggish, or even impotent. This positive association between vegetarianism and sex is endorsed through PETA's annual "sexiest vegetarian" contest, which receives over a million online hits (PETA Annual Review 2012).

Consumer Choice and Abundance

Through emphasizing choice, activists appeal to the myth of liberal individualism that romanticizes the self-directed individual, celebrating freedom of choice and a desire to have plentiful options. For example, COK's materials repeatedly empower people through use of the word *choice,* such as in asking consumers to "choose vegetarian" or "tryveg.com." PETA's videos also emphasize the consumer's ability to choose, framing those food choices as important, such as when the narrator in *Meat Your Meat* says, "Every time we eat we make a choice. Choose vegetarian," and the video exposing a Tyson slaughterhouse says, "We have a choice. We can choose cruelty to animals or we can choose compassion. Please go vegetarian." Even the use of the word "please" is a way of emphasizing that consumers have the power to choose and no force is involved. In rare cases, it is the nonhuman animals who are shown pleading with viewers to choose vegetarian, such as in a few of PETA's collateral materials where the farmed animals say "please don't eat us."

COK urges consumers to take advantage of their freedom of choice to help farmed animals by emphasizing that the animals have no freedom of choice over their own lives. COK's egg brochure and its pork leaflet both state that the animals, "don't have a choice – but you do." Similarly, COK has a television spot called "Choices" that asks, "Would you choose to live like this?" as it shows crated animals. Then it declares, "We have a choice. They don't." Also, In COK's FAQ section, in response to the common argument that other animals eat each other so we should eat animals too, COK provides the distinction that humans have the "choice" not to eat animals, as it is not necessary for human survival.

Choice also relates to the value of freedom in terms of not being bound by restrictions on food – even though this may seem ironic as most people associate veganism with *not* eating certain things (or, more specifically, I would say not eating *anyone* and their bodily products). Yet PETA's goveg.com "Veg101" section declares that vegetarians eat "whatever we want," which is an unusually liberating phrase that implies the choice to eat vegetarian foods is a satisfying preference and not a sacrifice. All animal organizations illustrate this abundance and variety of food choices with images of bountiful produce, packaged store-bought foods, and home-cooked meals. Even the occasional suggestions that consumers eat less meat, instead of eliminating meat, is another way activists emphasize that it is the *consumer's* choice to what extent they wish to change their diet. When viewing each animal organization's messages as a whole, however, it is clear that each group believes the *best choice* consumers could make would be to eliminate all animal products.

Enjoyment of Foods that are Pleasurable and Easy to Access

To showcase the value of pleasure and enjoyment, every animal group highlights the positive aspects of vegan foods, recognizing that taste, convenience, accessibility, and variety are very important to food consumers. To create a positive connotation with vegetarian foods, animal organizations often accompany messages with cheerful, bright colors such as green, yellow, and blue, connoting freshness, and showcasing photos of ripe produce, name-brand convenience products, and hearty cooked dishes of common favorites. Frequent use of words like "tasty" and "delicious" imply consumers won't compromise on taste. The recipe section of COK's starter guide, labeled "Recipes for Vegetarian Delights" assures readers "eating vegetarian foods doesn't mean giving up the tastes you love." PETA2's teen booklet labels its recipes, "Tantalize Your Taste Buds." And FARM also uses positive marketing when enticing readers to sign up for its Meatless Mondays campaign: "Have fun. Remember, going veg isn't about restricting your diet – it's about discovering new possibilities and experiencing fresh, exciting flavors." In this way, plant-based foods are framed as satisfying.

Additionally, to combat concerns that finding vegan foods will be a time-consuming inconvenience, the *ease* of the diet is often emphasized, especially by PETA, by stating that many accessible vegetarian options exist now. PETA's starter guide explains "restaurant options for vegetarian diners keep getting better and better," and "you can now find veggie burgers and other mock meats and soy milk in pretty much every supermarket nationwide, including Wal-Mart." Equally optimistic, COK and Farm Sanctuary also declare that it's "easier than ever" to go veg. COK's veg guide has sections titled "Simple and Delicious: Vegetarian Dining" and another is called "The Easy Way to Transition to Vegetarian Eating." And COK's vegetarian eating brochure contains a section called "It's as easy as 1-2-3." And Vegan Outreach's main booklets tell readers, "exploring a meatless diet is simple."

Belonging or Desire for Popularity

All organizations emphasize the growing popularity of vegetarianism, presumably so it does not seem like a fringe lifestyle or odd dietary choice. People do not want to be alienated, so, by emphasizing popularity, activists provide assurance that vegetarianism is socially validated. In actuality, some leaders admit in interviews that the percentage of vegetarians has not risen substantially over recent decades, but it is popular with certain demographics, such as women and youth. However, even if rates of vegetarianism are not growing rapidly nationwide, the sale of vegetarian foods is. Farm Sanctuary's starter guide assures readers there are a wide variety of people who eat vegetarian, saying "from former cattle ranchers to Hollywood celebrities, more and more people from every corner of America are recognizing that vegetarianism is good ..." and "after years on the fringe, meat-, egg-, and dairy-free fare has earned a well-deserved place in the American food culture." One page is dedicated to proving vegetarians are in "good company," as the "best people" have gone vegetarian for ethical reasons, showcasing famous influential leaders/artists throughout history who practiced vegetarianism, such as: Gandhi, Thoreau, Einstein, Harriet Beecher Stowe, and Cesar Chavez.

PETA often appeals to our desire to be part of the "it crowd." Celebrity headshots and quotations are used to demonstrate that vegetarians are morally progressive, healthy, attractive, and popular among a diverse demographic. The following famous actors and artists grace the cover of PETA's veg starter kit: Natalie Portman, Alicia Silverstone, Pamela Anderson, Common, Joaquin Phoenix, and Paul McCartney. PETA's teen booklet features attractive young stars under the headline "everyone's doing it." To further emphasize that beautiful people go vegetarian, PETA promotes its "sexiest vegetarians" annually and uses naked or scantily clothed bodies in some campaigns.

Pride in One's Moral Integrity by Living One's Values

I define moral integrity as consistently reflecting one's values through actions and applying those values uniformly, openly, and fairly in all situations. Most of the activists appeals were morally-based, indicating that people value their self image as a person of good moral character and integrity, which means even their consumption habits must be a true reflection of their values. In an appeal to the major, presumed value of compassion, animal activists activate the logic of moral consistency as such: if people already care about the welfare of cats and dogs and do not want to see them harmed, and if farmed animals are equally sentient, then it would make sense that compassionate people would not want to see farmed animals harmed either. To show consensus for farmed animal welfare values, vegetarian guides for Farm Sanctuary and PETA use survey data to prove that most Americans are in favor of legal protection of farmed animals and against intensive confinement. But consensus doesn't exist in favor of saving farmed animals from being killed for food. However, there seems to be consensus that people should not kill dogs and cats for food. So animal activists leverage this contradiction to direct attention to moral inconsistencies in American attitudes and treatment (killing) of non-human animals.

Messages by Farm Sanctuary, FARM, PETA, and COK use questions as a tool to provoke viewers to rationally justify why they *eat* certain species and *befriend* others, implying it is logically inconsistent decision who gets killed. A COK t-shirt shows a photo of a dog seated on a dinner plate with a knife and fork on either side of him.[4] The headline asks "Why not? You eat other animals, don't you? Go vegetarian." Similarly, a FARM vegetarian postcard shows a picture of a cat and a piglet nose to nose with the question "Which do you pet? Which do you eat? Why?" Farm Sanctuary uses this questioning technique the most. They have a t-shirt and other collateral materials with drawings of a happy dog and cat and an anxious cow and pig with the question, "If you love animals called pets, why do you eat animals called dinner?" The intentional use of the word *called* implies that humans treat nonhumans according to the arbitrary or socially constructed ways humans choose to define them, more so than how it has to be or naturally "is." The same phrasing and question is used by vegetarian actor Corey Feldman in a Farm Sanctuary PSA showing him petting a turkey and telling viewers that farmed animals have the same "emotions, personalities and intelligence" of the cats and dogs that are part of American families.

---

[4] I have this t-shirt and was pleased to meet this three-legged dog in person at a COK table at an animal rights conference.

And to help create empathy for sea animals, a PETA brochure says we humans wouldn't "stab our cat or dog through the mouth" (as a fishing analogy), and "none of us would drop a live cat or dog into boiling water. Why should it be any different for lobsters?" And to compare the act of eating fish to eating dogs, PETA often cites a quote from aquatic expert Dr. Sylvia Earle saying, "I wouldn't deliberately eat a grouper any more than I'd eat a cocker spaniel," based on their personalities.

America's animal cruelty laws are inconsistently applied between farmed animals and companion animals, as Farm Sanctuary, COK and PETA, inform us. Several Farm Sanctuary print materials simply say that farmed animals are excluded from most state anti-cruelty laws and from the federal Animal Welfare Act. COK's vegetarian guide says, "the animals who we eat are treated so abusively in this country that similar treatment of dogs or cats would be grounds for animal cruelty charges in all 50 states," and PETA's vegetarian guide claims that billions of animals are killed by the meat industry "in ways that would horrify any compassionate person and that would be illegal if cats or dogs were the victims."

Farm Sanctuary specifically uses the word "all" in places to emphasize how every animal species, including farmed animals, should be included in one's circle of compassion. Their sticker shows a calf and encourages us to, "extend compassion to *ALL* beings," and a t-shirt shows a piglet and reads, "*All* babies need love." Vegan Outreach also appeals to moral consistency, as their booklets openly talk about the need for people to widen their "circle of compassion" to include farmed animals. It states that most people are "appalled" by farm animal cruelty, not because they believe in "animal rights," but because they "believe animals feel pain and that morally decent human beings should try to prevent pain whenever possible." In this way, the utilitarian appeal is not asking for a *change* in values, since it assumes people are generally supportive of animal welfare, but rather it asks for a more equal application of this welfare value. In part, this emphasis on living one's values can be summarized in a quote PETA's veg guide uses from actress Natalie Portman to explain her moral reasons for becoming a strict vegetarian: "I just really, really love animals and I act on my values … I am really against cruelty [to] animals."

Animal organizations attempt to show that veganism is a true reflection of one's compassion, while supporting so-called "humane" farming is not, so they undermine the welfare claims made by some free-range farms. For example, Farm Sanctuary's position paper on "Humane Meats" says that people who are "sincere" in their concern for animals will stop eating them, insinuating that animal lovers are hypocritical if they still eat the flesh of animals, even "free-range" animals. The position paper goes on to suggest that veganism is a path to a "deeper level of compassion," which emphasizes its use as a tool for moral development.

Morality itself is based on notions of right and wrong. While overt use of the words "right" and "wrong" do not appear much in any of the organizations' messages, PETA does use them in several places. PETA's *Chew on This* DVD and television spot ends with the declaration, "you know this is *wrong*" when showing factory farm cruelty. PETA has a poster series and some leaflets with headlines that state, "What is *wrong* with…" meat, eggs, or dairy. On the poster, the solution is to "Join the Vegetarian Revolution," which is an unusual phrasing in the call-to-action that implies a moral uprising is warranted and is occurring. This moral uprising appears to be led by celebrities, so it is characterized as fun and trendy more than militant. More seriously, FARM refers to those who speak out for farmed animals on World Farm Animal Day as "people of conscience."

One of the few instances when the moral word "right" is used is in the beginning of PETA's *Vegetarian Starter Kit* saying vegetarianism is a way to "eat right" for health, animals, and the planet. Another time is in PETA's *Chew on This* DVD when it says, "might does not make right," a phrase also used once in COK's vegetarian guide. The PETA DVD utilizes other moral language to promote vegetarianism by also saying there's "no excuse for murder," as these animals do not want or deserve to die, and that this "isn't fair." The use of the word *fair* is uncommon, as animal rights organizations rarely made direct appeals to *fairness* and *justice* for nonhumans.

My Evaluation of the Implications of Select Values-Based Appeals

So far in this chapter I examined how activist food advocacy messages positioned the human subject in terms of what values humans are said to possess or should possess that might motivate them to concur with the activists' proposed solutions. In summary, to demonstrate veganism's fit with consumer identities, animal rights organizations most prominently appealed to our presumed values of compassion, respect for animals as subjects (not objects), healthfulness, environmentalism and moral consistency. They also appealed to other values such as: desire to improve the world and make a difference, choice, pleasurable and convenient food, social belonging and appeal, life, concern for fellow human beings, honesty, populism, naturalness, freedom, and patriotism. In keeping with suggested framing and rhetorical strategy for social movements (Benford and Snow 2000; Ryan 1991; Therborn 1980; Zald 1996), these values generally fit within American cultural norms, which should enhance their resonance.

I particularly want to ascertain if and to what extent animal activists promoted values that were representative of animal *rights* ideology, not just welfare. While I think the values that they promoted did not usually conflict with or contradict animal rights values, only certain values actually *promoted* an animal rights viewpoint. And these values only did so when animal activists specifically framed them in ways that created this connection. Values that were either informed by or promoted an animal rights ideology to some extent were: compassion, respect for sentience and individuality, moral integrity and consistency, honesty, life, freedom, naturalness, belonging, and desire to make a difference. In the following sections, I discuss how each value was or could be directly related to animal rights and how this framing fits with communication strategy recommendations from scholars.

Compassion

While the notion of compassion resonates with popular sentiment because it connotes concern for the welfare of other animals, the animal rights organizations optimistically implied, but did not overtly state, that people's compassion for animal welfare will extend to an animal rights perspective that animals should not be exploited or killed. For example, according to activists' frames, the problem that rouses compassion is viewing factory farm cruelty (a welfare frame), but activists then associated this compassion with leading people to a vegan solution (rights frame) and not with switching to less cruel animal products. So, by this association, animal organizations applied a deeper or stricter meaning to the concept of compassion and suggested that to be *truly* compassionate is to avoid killing or exploiting any animals for food. I see this as an example of frame amplification (Snow et al. 1986).

The value that I believe is more directly related to the principle prohibiting farmed animal killing and exploitation is *justice* instead of compassion. While compassion is a necessary first step or component of justice, as it creates an initial concern and caring for other sentient beings, it is not as fundamental or direct as justice is at implying that these animals have a right to live free of human control. Activists' appeals to compassion tended to be more restricted to the notion of avoiding suffering and cruelty. So they condemned certain poor husbandry practices more than they condemned the entire practice or concept of animal farming.

Respect for Sentience and Individuality of Other Animals

Appeals to compassion generally did not ask the audience to think differently of farmed animals, as these appeals relied on audience members *already* viewing farmed animals as sentient beings who are capable of feeling pain. As the activists often framed cruelty as the main problem, these appeals to compassion implied that the public currently must have some level of respect for farmed animal sentience, at least related to their ability to feel pain and suffering. But I think the value of respect for sentience can be framed to be more transformational in favor of animal rights if the frames convince people that farmed animals not only feel pain but are individuals who have emotions, consciousness, and unique personalities. Then someone's respect for farmed animal sentience might presumably deepen into seeing them as individuals who have the same right not to be eaten as do other individuals, such as humans or horses, based on the fact that Americans generally view the latter as individual *subjects* not objects. Activist organizations, particularly Farm Sanctuary, PETA, and Compassion Over Killing, often did promote the idea that farmed animals were unique, thinking, feeling individuals.

Moral Integrity and Consistency

Directly related to valuing farmed animals as fellow "subjects of a life" (Regan 2004, 185), is the desire to value one's own moral integrity by respecting the rights of these fellow subjects in a consistent manner. Activists' frequent use of dog and cat comparisons with farmed animals makes logical and moral sense as a way to encourage people to question the irrationality of the status quo's unjustified categorization of some animals into subjects and some into objects. Statements such as PETA's "If you wouldn't eat your dog, you shouldn't eat fish," are an important first step in getting people to acknowledge that they avoid eating *certain* animals for moral reasons, so they should also consider the moral inconsistency of their eating *any* animal.

But the use of the moral consistency value has limitations similar to the value of respect for sentience if all the comparisons are restricted to being between farmed animals and other domesticated *nonhuman* animals, since this tends to draw mainly upon a desire for consistent application or expansion of current animal *welfare* values only. Animal rights organizations are not taking the opportunity to promote animal *rights* values that would challenge the human/animal dualism and compare the subject status of farmed animals to that of human animals. Part of an appeal to the value of moral consistency could be to ask people to consistently apply some of their values of justice and rights, as typically directed mainly at *humans*, to nonhuman animals, namely domesticated animals whom we have tended to use for food. These human rights values include the right not to be exploited or enslaved and the right not to be killed (if one is not guilty of any violent crime). As women and people of color have a history of being discriminated against on the basis of their association with lowly nonhuman animals (Adams 1990; Spiegel 1996), animal activists could more frequently draw upon ideas of moral progress and human social justice analogies, abstracting them where appropriate to fit nonhuman animals, as do Regan (2003), Singer (1990), and Francione (1996). I believe the process of expanding the idea of basic human rights and applying it to the treatment of other animals is an example of frame transformation (Snow et al. 1986).[5] But activist messages in this study

[5] To clarify the boundaries of this analogy, animal rights goals specifically ask for humans to apply human rights values to nonhuman animals, just to the point of preventing humans from domesticating, exploiting, and needlessly killing fellow animals. But it does not dictate that humans interfere with the natural predation cycle followed between groups of wild nonhuman animals, according to their society's culture or ecological principles found in nature. Basically, animal rights guides the behavior of humans not nonhumans.

rarely drew upon transformational human rights analogies, instead frequently relying more on bridging people's desires not to eat pets to a desire not to eat farmed animals.

When one considers how a moral integrity appeal aimed at mitigating *human rights* abuses would not be framed as a consumer choice issue but rather as an issue of civic obligation to uphold rights, I believe that, similarly, the moral consistency appeals for extending rights to nonhuman animal subjects should be complemented with an "ethical obligation" frame more so than the "choice" frame that animal organizations tended to emphasize. While ethical veganism *is* a choice, as it is not illegal to eat farmed animals, it is better aligned philosophically with a justice frame than it is with a consumer choice frame, as the latter may make veganism another trendy lifestyle choice instead of an ethical obligation (Maurer 2002).

Honesty

A subset of moral integrity is an appeal to honesty. Activists did so but largely in terms of appealing to consumers' desire for honest marketing tactics from agribusiness, being upfront about their corporate practices. But within a moral integrity frame, activists should more frequently appeal to consumers being honest with themselves, by acknowledging their own role in agribusiness problems. All problem frames that blamed the consumer implied that consumers needed to take an honest look at the "reality" of factory farm cruelty and environmental destruction that agribusiness hides from them.

PETA also used an isolated, but useful, honesty frame telling parents they should not have to lie to their kids about where their food comes from. This implied that adults are ashamed of the killing of animals for food and know it would emotionally upset children, so they remain complicit in hiding the violence they financially and tacitly support. An honesty frame could state that one should willingly, openly, and frequently confront the agricultural practices and consequences behind one's food choices to ensure they are in accordance with one's own values in order to maintain moral integrity and model that value for one's children.

Life

Every animal rights organization, with the exception of Vegan Outreach, alluded to valuing life. I believe life to be a central component of animal rights, as in other animals having the right not to have their lives taken by humans, except in self-defense. The right of all animals to their own life is more central to animal rights ideology than is the right to be treated well in captivity, so I believe that animal activists should more frequently appeal to people's value for protecting and maintaining life over death, instead of primarily appealing to compassion for suffering. Additionally, times when activists portrayed farmed animals requesting and desiring that we save their lives also complemented the previous value of respecting animal sentience, as it imbues them with agency. The idea is that humans should value the life of anyone else who has the ability to value his or her own life too. I like how Mary Midgley (2008) put it when asked why animals matter, "because things matter to them" (21). This allows animal rights organizations to draw comparisons that can break down the false human/animal dichotomy by showing how all animals, human and nonhuman, value their own lives. Farm Sanctuary used this approach on the t-shirt featuring Buddha's quote, "All beings tremble before violence. All fear death. All love life."

Freedom

Besides Farm Sanctuary saying its sanctuary residents enjoyed freedom and PETA once saying, "everyone wants to be free," the value of freedom was implied more than it was explicitly mentioned. For example, activists emphasized and problematized the confinement, crowding, immobilization, lack of space, and lack of outdoor access on factory farms. They contrasted this immobility with visuals showing the relative freedom of rescued farmed animals outdoors in a sanctuary setting reminiscent of a small family farm and with the few images of "food" animal species in the wild, especially fish. One of the problems is that these visuals set up an implied definition of freedom that does not mean freedom from domestication and farming but simply freedom from indoor, intensive confinement. This tacitly supports small, less cruel, free-range farming rather than supporting the idea that nonhumans should own their own bodies and be free to control their own lives and families. Some organizations, such as PETA, got closest to this latter frame through occasionally discussing the rights of dairy cows and egg-laying hens to own their own offspring, milk, and eggs.

Related to freedom is the notion of control over one's body and choices. This was emphasized by some of Compassion Over Killing's messages that stated humans have a choice but these farmed animals do not. This implies that the farmed animals are stuck in a bad situation through no fault of their own, but humans have the freedom to choose whether or not to free them from this bad situation. This frame could be more explicitly tied to the nonhuman's freedom and related notions of choice and opportunity if it explained that while *wild* animals often have the opportunity to escape being eaten by predators, *farmed* animals are given no such opportunity to avoid becoming prey, in this case to a human predator. Then the frame emphasizing the lack of freedom and forced captivity can link up with *natural* principles of freedom and "survival of the fittest" as well as American *cultural* principles supporting justice, fairness, and opportunity.

I argue that appeals to the value of American pride should align with notions of freedom rather than the activists' tendency to appeal to American pride based on a somewhat humanist and elitist idea of America's advanced civilization and "humanity." Freedom and liberty are positive principles that are heavily associated with the rhetoric of America. America's Declaration of Independence proclaims everyone's rights to "life, liberty and the pursuit of happiness," the national anthem declares America to be "the land of the free," and the pledge of allegiance claims America provides "liberty and justice for all." Therefore, activist messages aimed at Americans could cite freedom terminology more explicitly to align the animal rights movement with accepted democratic principles that resonate with the American public and are commonly part of rights movements (Bormann 1971; Campbell 1989). This frame amplifies or transforms the idea of having the right to freedom so it applies to other animals in order that they may seek their own versions of life, liberty, and the pursuit of happiness, free from exploitation. While this freedom frame is more logically applied to "wild" nonhumans who are commercially caught for food, particularly fish, it could also be used to emphasize the animal rights principle that it is not in anyone's best interest to be domesticated and kept in captivity.

Freedom is a complicated frame in the case of farmed animals who have been selectively bred in captivity and are no longer equipped to survive in the wild. Frames could explain the vision that nonhuman animals should have basic freedom over the right to their life and to own their own body, milk, eggs, and offspring, as all wild animals do. But this freedom frame would then need to clarify that the animal rights movement generally does not intend that all existing farmed animals should be set free into the woods, but rather, the movement seeks an end to their intentional breeding and use by humans. It should be admitted that this would likely result in the eventual extinction of most highly-domesticated farmed animal breeds who no longer have a place in the ecosystem and who are more adapted to rapidly and painfully growing unnaturally fat than they are to living in a natural environment.

Naturalness

Freedom also ties into the value of naturalness, as other animals in nature do not farm other species in captivity as a food resource. So while animal activists' frames generally appealed to naturalness by framing *modern* agribusiness practices as "unnatural" in comparison to traditional animal farming, animal activists could extend a naturalness frame out to communicate that *any farming* of animals for food is itself unnatural when viewed in relation to common forms of predation in nature. However, this frame is complicated by the fact that humans have practiced animal agriculture for thousands of years, so a counterargument may be that farming has become naturalized for our species and those whom we have domesticated, as well as the fact that plant agriculture is largely unnatural too (Pollan 2006). To counter this, I suggest that appeals to animal agriculture's unnaturalness are best done within a meta-frame of freedom or justice.

In addition, activist appeals to the naturalness of a plant-based diet for human nutrition were associated with the value of health. At some point in any vegetarian argument, activists must use a health frame to validate the diet. A health argument is useful to demonstrate that predation is *unnecessary* to human survival in cases where a variety of plant-based proteins are available. Proof that animal products are unnecessary to human health in America is integral to supporting the frame that problematizes the killing of animals for human food.

For honesty in communication, animal activists should take care not to make health or naturalness claims that go beyond what can be well substantiated, and those references should be clearly cited. It was not the goal of this book to assess the accuracy of claims made by animal organizations, but I did discover in the interviews with activist leaders that no one took an extreme utilitarian position of stating they would intentionally misrepresent the facts to achieve animal rights goals. Erica Meier said Compassion Over Killing is not a nutrition organization and prefers to refer people to other sources if they want more health information, which seems like an honest way for an animal rights organization to approach the topic.

Vegan Outreach's co-founder and President Jack Norris actually is a registered dietician. With their *Guide to Cruelty-Free Eating,* Vegan Outreach is perhaps the most careful and conservative animal rights organization in addressing the health topic and explaining any issues, not glossing over potentially negative information. Because of industrialized agriculture, people do not obtain their food as their ancestors did in the wild, so a modern, solely plant-based diet presents a few potential issues of nutrient deficiency, such as vitamin B-12, of which potential vegans should be made aware (Mason 1997; Melina, Davis, and Harrison 1995). These issues can be overcome with supplementation and planning (Academy of Nutrition & Dietetics 2009; Melina, Davis, and Harrison 1995), but they should be openly addressed along with explanations of why these modern issues exist, so the naturalness frame is still supported. To foster honest communication, if activists are going to point out any potential risks with modern *animal*-based products or diets, they should be willing to point out any potential risks with a modern, solely *plant*-based diet.

Providing basic, accurate nutrition details to aspiring vegans, or pointing them to reliable health sources, is part of a commitment to communication ethics that builds credibility for animal rights and shows a related concern for the human animal. In fact Bruce Friedrich stated that PETA cares about humans too as fellow animals, so he did not think there was any conflict with an animal organization focusing on human health, implying it wasn't just utilitarian. But other animal organizations tended to admit a more utilitarian motivation in using the health frame for largely strategic reasons, as people were naturally motivated, at least in part, by self-interest. Additionally, Erica Meier's emphasis on providing *tools* that could keep people healthy and happy on a long-term vegan diet is another way that the health or naturalness frame can be utilitarian.

Belonging and Social Appeal

When organizations such as PETA and Farm Sanctuary show the diversity of people who are going vegetarian and let the public know there are millions of people choosing this diet, it provides validation that this is not just a subculture of alternative youth or older hippies. This practice of emphasizing diversity fits with Tarrow's (1998) framing challenge to avoid creating a narrow or elitist identity when attempting to build unity and attract people to the movement, as well as Cooney's (2011) recommendations that activists recognize that people have a strong desire to conform and fit in to what is socially acceptable. Through PETA's use of celebrities, and Farm Sanctuary also using moral leaders in its vegetarian guide, they built a concept of unity or identity based not on age, race, gender, or style, but on people holding similar values and acting with integrity. This fits with Crompton and Kasser's (2009) recommendations to focus on promoting intrinsic values. However, while there was diversity, the images still favored attractive, younger, white people.

I found that Farm Sanctuary, especially, and PETA to an extent, did a good job in emphasizing the moral reasoning behind the celebrities' and leaders' choices to be vegetarian so that the focus was put on animal ethics as the unifying rationale. Farm Sanctuary's use of the theme that you are in "good company" highlights belonging to or emulating a group of people who have good virtues, rather than just a group of people who are well-known or physically attractive. To maintain focus on the nonhuman animals and to avoid making vegetarianism look like a Hollywood fad, it is important to include leaders from throughout various stages of history and different cultures to ground ethical vegetarianism in a long history of moral thought, in keeping with communication scholar recommendations to make historical connections between ideas (Ryan 1991; Therborn 1980). While most of these historical leaders were men, this can serve the utilitarian purpose of helping a male audience recognize that vegetarianism does not have to be seen as an effeminate dietary choice. Additionally, I think animal activists should feature stories from former animal farmers and hunters to help provide further diversity in terms of masculine and rural perspectives and to dispel any ideas that vegetarianism is just an urban alternative subculture.

Desire to Make a Difference and Be Significant

The previous theme of identifying and emulating leaders who practice ethical vegetarianism is related to the idea that through vegetarianism you as an individual can also make a difference and do something important. Animal organizations did not draw that exact parallel between the reader being as important to social causes as Gandhi or the Dalai Lama, but they sometimes did say that vegetarianism was the "most important" step a person could take everyday to solve a variety of problems caused by animal agribusiness. To provide the motivational aspect of framing, activists emphasized the "power" every person has to "make a difference" by saving animal lives, mitigating global warming, and protecting the planet. These altruistic messages empower the average American to be an activist at every meal, without any more effort than simply eating plant-based foods. This turns the mundane *private* act of eating, done mostly for pleasure and sustenance, into a *public* act of more social, moral, and political significance.

In support of both means-oriented and ends-oriented communication perspectives, framing veganism as a vehicle for personal growth helps to emphasize the moral and altruistic rationales for veganism and create a stronger identity and commitment than messages that emphasize self-interest (Crompton and Kasser 2009; Maurer 2002). Yet, being proud of oneself and feeling consequential is also in one's own self-interest. While veganism is based on the belief that the lives of *other* animals matter, it also helps the vegan believe that his/her *own* life matters too by recognizing that his/her food choices have far-reaching impacts.

IN CONCLUSION

The animal rights movement seeks a broad identity for veganism if it is to become a mainstream dietary choice. Perhaps that is why the appeals seem bipartisan, non-gender-specific, and non-denominational. Activists attempted to show how a variety of values with which many Americans already identify are a natural fit for choosing plant-based foods and are incongruent with consuming animal products. According to activists' representations, plant-based diets embody altruism and caring, health, humanitarianism, and environmental responsibility. This could imply that America's dietary choice to eat animal meat, eggs, and dairy represents the opposite – selfishness, irresponsibility, unsustainability, violence, injustice, and apathy toward the lives and suffering of humans and other animals. Certainly if there is any unsavory character in this vegetarian discourse, it is agribusiness, not consumers; you don't see the animal rights organizations constructing a heartless meat-eater as much as they choose to construct a thoughtful future vegan. Taken as a whole, the collective action frames' subtext reads, "Now that you know about the problems with animal agribusiness, an ethical and rational person such as yourself will surely make the right choice and go vegan." But once consumers *do* know about (or finally admit they know about) the suffering, injustice, pollution, etc., and if they choose to continue to support it, even at a diminished level, how is the animal rights movement supposed to address them positively as potential heroes then? At this stage in the movement, we are still assuming that consumers cannot live with being hypocritical and will therefore agree to extend welfare values (or hopefully rights values) out to farmed animals. In both the current animal rights discourse and my recommended rights-based framing, a lot is riding on a belief in humanity's compassionate tendencies and their psychological need to view themselves as people of moral integrity.

So, do animal rights messages tend to cultivate these needed compassionate and altruistic tendencies or do they assume humans are mainly self-interested? The next chapter answers this question, continuing an analysis of values.

Chapter 6

Appealing to Altruism or Self-Interest?

Like other social movements, animal rights is a movement for moral progress that asks society to show more respect, compassion, and fairness toward *others* – especially those who have been "othered" and marginalized by the dominant group. While those humans who are marginalized can sometimes participate in a social movement on their own behalf, animal rights is truly an *other-directed* movement in that it relies solely on human volunteers to advocate on behalf of nonhuman animals.[1] Since the promotion of altruism, caring, and concern for others is so key to widespread adoption of animal rights, I wanted to identify the prominence that animal rights organizations place on promoting selfless values such as altruism, caring, justice and social responsibility toward *others* as opposed to more self-interested values such as personal gain. Within this question I also wondered to what extent animal activists' altruistic appeals toward others emphasize a concern for the *nonhuman* animal other (both farmed and free/wild), instead of only other humans, as the animal organizations' primary purpose is the protection of nonhuman animals. When activists appeal

---

[1] I do not mean to indicate that nonhuman animals don't have a voice or a perspective on their treatment and their lives, nor to deny the morality practiced by social animal species. Rather I mean that in the human legal system and among human society, humans naturally must take the lead negotiating how we could improve our treatment of nonhumans. In advocating for this change, activists must draw upon what they know about nonhuman animal personalities, capabilities, sensitivities, and preferences. And we can share their voice by showing their reactions to their captivity and our treatment of them. This is why it's so important for society to hear and see other animals speaking for themselves in audio-visual media – such as that obtained via nature documentaries, home videos, and undercover footage (see Freeman, Bekoff, and Bexell 2011).

to the audience member's concern for nonhuman animals, I refer to these appeals as "animal-centered" to differentiate them from anthropocentric appeals that are geared primarily toward a concern for the human animal.

The opposite side of the coin to promoting altruistic values is a focus on the audience's own self-interested values (so they can see what's in it for them personally), realizing that self-interest and altruism are not always mutually exclusive categories, as I will later explain. When discussing how veganism can fulfill one's own self interest, the focus is usually on how plant-based diets improve human health (via nutrition, disease-prevention, and weight-loss), reduce environmental risks to health and public safety, and offer satisfaction, convenience, and community.

As part of my analysis, I examined appeals to altruism and self-interest in the vegan advocacy materials of PETA, Farm Sanctuary, Farm Animal Rights Movement (FARM), Compassion Over Killing (COK), and Vegan Outreach in 2008. To summarize my research findings, all five animal organizations place an emphasis on promoting an altruistic concern for farmed animals (whether it be their welfare or their rights), primarily through dedicating a lot of space to animal issues and using farmed animal photos throughout messages. Relative to each organization in this study, Vegan Outreach and COK place the largest proportion of emphasis on nonhuman animal issues. They are also the smallest groups in the study, and the leaders informed me in their interview that their limited resources are directed more toward educating the public about farm animal cruelty rather than nutrition or environmental issues. Conversely, for pragmatic reasons, Alex Hershaft of FARM admitted that FARM mainly promotes vegetarian foods based on human self-interest values rather than animal rights (yet, I found that FARM's materials still emphasize farmed animals quite a bit). Farm Sanctuary and PETA, being the organizations with the largest resources in my study, produce the greatest quantity and array of materials. Therefore, they have the space to branch out into more anthropocentric issues; yet they both, especially Farm Sanctuary, still spend the majority of their space and efforts promoting animal-centered appeals (ex: cruelty) rather than anthropocentric appeals (ex: food, health, or environment).

In this chapter I explore how each animal organization's appeals focus on:

- Altruism toward nonhumans versus human self interest;
- Anthropocentric altruism (caring for fellow humans);
- Environmental appeals as both selfless and self-centered;
- The connection between morality and mental health (where selflessness and self-interest overlap).

Throughout, I discuss the implications of these choices, advocating for campaigns promoting animal-centered altruism.

ANIMAL-CENTERED VERSUS ANTHROPOCENTRIC APPEALS

While all animal organizations promote veganism as a solution to proposed problems with meat and agribusiness, it is important to ask upon what basis is this veganism promoted. To investigate, I primarily examine the vegetarian starter guides and "why vegetarian?" sections of activist websites as the most pertinent sites for assessing the emphasis that these five animal organizations place on the audience member's self-interested values versus altruistic values. In this section, I discuss each organization in relative order of their emphasis on animal-centered appeals over anthropocentric appeals.

Vegan Outreach

Vegan Outreach's three main booklets used for leafleting are dedicated to farmed animal cruelty and compassionate messages –*Why Vegan?*, *Even if You Like Meat*, and *Try Vegetarian*. Each booklet's cover features photos of farmed animals only and uses the word "suffering" or "cruelty," which conveys that respect for nonhuman welfare is the main reason to give up (or reduce) eating meat. And veganoutreach.org also has chickens across its header. Approximately 13 of the 16 pages in *Why Vegan*? and *Even if You Like Meat* are focused on nonhuman animals, placing the self-interested health and food-oriented pages toward the back. The *Try Vegetarian* booklet takes a more self-interested approach, relatively speaking, by starting out talking about health, with only half the pages dedicated to nonhuman animal altruism. However, photos of farmed animals feature prominently on almost all its pages. Vegan Outreach does have a health section on its website and a separate *Guide to Cruelty-Free Eating* that is dedicated more to health and practical food preparation issues than are its three main leafleting booklets. These nutrition resources are there for the newly converted to seek out once they are convinced they don't want to contribute to animal suffering.

Compassion Over Killing (COK)

With compassion built into their name, perhaps it's fitting that all of COK's campaigns, video footage, print pieces, and television spots are dedicated to the altruistic purpose of exposing the public to the harsh realities of factory farm cruelty. Their website features photos of farmed animals on the header of all pages. COK's veg eating brochure has a piglet on the cover and starts with animal-centered reasons to go vegetarian before proceeding to health and environmental reasons.

COK's *Vegetarian Starter Guide* dedicates a quarter of its pages to farmed animals, with the six-page animal section coming after the three-page health section, followed by a two-page environmental section. The guide has photos of farmed animals on its cover along with fruits and vegetables. Over a third of the booklet's pages simply help people make the transition to vegetarianism with recipes and shopping tips.[2] Erica Meier told me that COK privileges the issue of farmed animal suffering but also felt it was necessary for COK to provide people with the tools they need to go vegetarian and maintain that lifestyle, hence the pages dedicated toward food procurement and recipes. Also in this latter effort, COK has launched a series of city-specific vegetarian websites, providing tips on eating vegetarian in major cities such as Washington, DC and Portland, Oregon.

Farm Sanctuary

Farm Sanctuary is heavily focused on farmed animals, as it is the only organization in my sample that actively rescues and provides homes for animals. Farmsanctuary.org showcases a huge photo of sanctuary residents across the top and also portrays at least one photo of a

---

[2] This is technically self-interested, but I perceive it to be more value-neutral than the health section, as preparation and access to food is more just a practical concern and not a rationale for going vegetarian.

farmed animal on each webpage. All of their campaigns, advertisements, videos, and the vast majority of print pieces and collateral materials are dedicated to farmed animals.

They have a leaflet called *20 Reasons to Go Veg for Life* which is evenly divided between animal, environment, and health reasons; but the animal-centered reasons do get some premier placement as the first and last reasons listed, and the majority of photos include farmed animals. Of all the reasons listed, approximately half could be considered altruistic, including all the animal-oriented reasons and many environmental reasons. Farm Sanctuary also has a lengthy *Guide to Veg Living*, one quarter of which is dedicated to altruism toward nonhuman animals, with cows featured prominently on the cover as the only photo. Pictures of named farmed animals (such as Rudy and Charlotte), sometimes shown with human companions, appear throughout the guide. The guide opens with a page of well-known people talking about their moral reasons for going vegetarian. This is followed by four pages on health before a three-page section on farmed animals, followed by a two-page environmental section, which includes a "ransacked oceans" paragraph on sea animals. Similar to COK's and PETA's vegetarian guides, one third of the guide is dedicated to practical information such as recipes, food products, and tips on making the transition.

People for the Ethical Treatment of Animals (PETA)

PETA uses the words "ethics" and "animals" in their name, so it is appropriate that they should focus heavily on nonhuman animals. While that is largely the case, PETA probably appeals to human self-interest slightly more often than do most animal organizations in this study, with the possible exception of FARM. Anthropocentrism is evidenced, in part, by PETA's greater use of celebrities and photos of humans, along with their greater emphasis on the value of attractiveness (including sex appeal and weight-loss). These generally fall into the self-interested values category of "belonging and social appeal" (as discussed in chapter five).

PETA's *Vegetarian Starter Kit* resembles that of COK's and Farm Sanctuary's, with a similar proportion of pages dedicated toward nonhuman animal altruism in the center (approximately one quarter), another quarter dedicated to human health, and virtually the rest (almost half) covering food issues like recipes and tips on transitioning. The difference is that environmentalism is reduced to half a page, and more emphasis is placed on fitness (like athleticism and weight-loss). Also, the cover page of PETA's starter kit features photos of celebrities and food but never features an image of a nonhuman animal. A farmed animal does not appear in the kit until page five, while the first few pages feature nineteen photos of celebrities. However, the majority of quotes by those celebrities deal with altruistic values toward nonhumans.

In PETA's "Top 10 reasons to go vegetarian in 2008," listed on goveg.com, over half the reasons are self-interested (with "slim down" being the first) and only one third of the reasons mention nonhuman animals. But PETA's popular *Chew on This* DVD lists thirty reasons to go vegetarian, and nearly two-thirds focus on farmed animal altruism, especially in the last half of the video. In the middle of the video, only two reasons focus on the environment and two on altruism toward humans. A quarter of the reasons are self-interested (health), and they hold a prominent place, dominating the first third of the video. However, the visuals overwhelmingly emphasize farmed animals, as opposed to humans, throughout. The last non-anthropocentric reason viewers are left to ponder is heavily moralistic – "because you know this is wrong."

Overall, PETA's numerous food collateral materials and print pieces are almost solely dedicated to farmed animal altruism, even though PETA still follows a trend of featuring celebrities heavily in print. And their extensive goveg.com site lists "cruelty to animals" and "amazing animals" as the first and second link under the "Why Vegetarian?" section. Only the health link is wholly self-interested, while links on the environment, world hunger, worker rights, communities, and government negligence are largely altruistic (even if more human-focused).

Farm Animal Rights Movement (FARM)

FARM is campaign-oriented rather than just having an overall "go veg" theme. Of the campaigns I analyzed, three are heavily altruistic (Bite Global Warming, Gentle Thanksgiving, and World Farm Animal Day) with a fourth campaign (Well-Fed World) emphasizing anthropocentric altruism (concern for humanity). Two are primarily self-interested (The Great American Meatout and Meatout Mondays). In talking to Alex of FARM, he seems to emphasize those Meatout campaigns as being most important, as he thinks people are most influenced by self-interest. Yet, in FARM's Meatout messages, even though they privilege self-interest (health and general food preferences), they do always mention farmed animals and environmental issues, to a lesser degree, at the end.

The "Why Vegetarian" section of farmusa.org lists issues in order of their perceived interest to the public. The list goes from most anthropocentric and self interested (health), to anthropocentric altruism (world hunger), to altruism mixed with self-interest (environment), to nonhuman animal altruism (farmed animals). The introduction includes a statement that reflects FARM's dual emphasis on self-interest and selflessness, "Although most people are motivated by health concerns, it is important to realize that dietary choices have much broader implications for planetary survival." In the introduction, they blame animal agriculture for its role in causing the "biggest problems facing America and the rest of our planet," namely "disease, hunger, environmental devastation, and death."

Anthropocentric Altruism

I discussed appeals to anthropocentric altruism in the previous chapter, in the sections covering the values of humanitarian concern for fellow human beings and populist values of fairness to the working class. To summarize, of all animal organizations, PETA puts the most emphasis on altruism toward other humans in its extensive goveg.com site that discusses rural communities, workers, and human hunger. But these issues are not highlighted elsewhere, such as in PETA's print pieces. Farm Sanctuary has a small section on rural communities in the factory farming section of its webpage and occasionally mentions hunger and worker issues in other materials. FARM is the only organization to dedicate a whole campaign to human hunger, particularly in less industrialized countries, but the campaign does not have a domestic focus on rural communities or worker issues. When considering all animal rights messages as a whole, anthropocentric altruism is dwarfed in comparison to the emphasis on altruism toward nonhuman animals and even, to a lesser degree, anthropocentric *self*-interest.

ENVIRONMENTAL MESSAGES, BOTH ALTRUISTIC AND SELF-INTERESTED

Each animal organization includes messages dedicated to protecting the environment, with Vegan Outreach using this appeal the least. I consider environmental values as both self-interested and altruistic because of human's ecological interdependence with the natural world for survival and the fact that issues like climate change affect all animal life. The question is: which is emphasized more in environmental messages of animal organizations, human self-interest or altruism? For example, when messages focus on the wellbeing of nonhuman species, such as wild animals, rainforests, or oceans, this is more altruistic. But when messages focus on domestic pollution and its human health risks, I categorize these as more self-interested. While it is an inexact science to separate these mixed messages into two distinct categories, my overall assessment is that animal activist environmental messages are both self-interested and altruistic but might lean more toward altruism.

PETA's environmental messages imply altruism when they suggest people should not eat animal products because it causes so much waste, inefficiency, and pollution. For example, the *Chop Chop* leaflet focuses as much on inefficiency/waste as it does on pollution and includes a section on the destruction of ocean life and deforestation. Also, the "what you can do" section PETA's goveg.com has this altruistic message, among others: "Switching to a vegetarian diet reduces your 'ecological footprint,' allowing you to tread lightly on the planet and be compassionate to its inhabitants." However, PETA's teen booklet appeals more to self-interest by placing a visual emphasis on human health hazards, featuring toxic icons, a polluted stream, and a barren landscape. It also shows a gas pump, which signifies expense and security in today's political climate.

Almost all of FARM's print materials briefly mention environmental protection. Plus, FARM has a dedicated environmental campaign built around global warming that also addresses all aspects of environmental devastation. The campaign includes a poster, t-shirt, postcard, and online section. All feature the earth logo, which humbles humans and emphasizes their mutual earthling status with all other living beings who share our planet. While these materials contain anthropocentric messages (some self-interested), particularly around pollution issues and the effects of global warming, they do mention animal-centered messages about the protection of ecosystems and wildlife. For example, the poster says that meat production "kills more wildlife than all other activities combined." The online, "What You Can Do," section includes the altruistic statement that viewers should go veg for "the Earth and ALL its inhabitants."

Farm Sanctuary's environmental messages are altruistic but include self-interested values slightly more often than other animal organizations, mainly because their environmental section includes some of the local worker and community health risks. Toward anthropocentric appeals, Farm Sanctuary's environmental brochure emphasizes pollution, mentioning "tainting drinking water" and the "health threats" and "respiratory problems" of air pollution, particularly for people living near factory farms. The brochure further emphasizes public health risks by showing pictures of medicine and a chemical plant in conjunction with a paragraph on "toxic drug residues" in meat. However, toward animal-centered appeals, it includes a comment on how these chemicals also put wildlife populations at risk, and it discusses dead zones next to a photo of a wild fish kill. The paragraph on "leaking lagoons" explains how cesspool leaks often sicken both humans *and* "native animals and plants."

Farm Sanctuary's environmental section of their vegetarian guide emphasizes the urgency of the need for dietary change based on the self-interested reason of protecting *our* lives and resources, stating that otherwise "the valuable resources on which our lives depend will continue to be eroded, depleted and polluted beyond repair," but most of the messages following this highlight risks to *both* human and nonhuman populations. And toward animal-centered altruism, the "ruin on the range" paragraph includes threats to endangered species and the killing of "wild animals" by the U.S. government's wildlife services division to protect ranching interests. The paragraph on "ransacked oceans" also mentions aquatic species' extinction, how aquaculture damages ecosystems, and how commercial fishing kills so many "bycatch"[3] animals (emphasizing the deaths of mammals and birds over fish).

---

[3] *Bycatch* is the term used for any unintended animals injured or killed by the fishing industry, meaning the seabirds, dolphins, turtles, or fish (deemed less edible or profitable) who get caught in nets/lines but whose flesh is not sold (these individuals are left to die and their bodies discarded). Bycatch is very common to fishing, meaning

The environmental section of COK's vegetarian guide is evenly split between human self-interest (pollution and toxins) and altruism (efficient use of resources and ocean biodiversity). Toward self-interest, it says the air and water we use are polluted and ends by saying humans should protect the planet "for ourselves and our loved ones." In favor of altruism toward nonhuman life, the section titled "Saving the Earth" shows a clear-cut forest, a bee on a flower, and a man trying to free a giant tuna caught in a driftnet. Additionally, in the paragraph on fishing, it explains, "local ecosystems are destroyed, devastating animals and plants."

MENTAL HEALTH AND MORALITY: HOW SELF-INTEREST OVERLAPS WITH ALTRUISM

The idea of altruism is not always devoid of self-interest, as was mentioned by several executive directors in my interviews with them in 2008. For example, Matt Ball of Vegan Outreach called the choice between self-interested and altruistic appeals a "false choice" because, by appealing to people's ethics, you *are* appealing to people's own self-interest. He said that being vegetarian:

> adds to people. It's not about "I've given up meat. I've given up cheese. I've given up eggs, and I suffer through the day because I don't have these things." It's really something that can be a very positive thing for an individual. It can really add to the meaning of their life – to their ethical satisfaction – to their fulfillment as a person.

Notice how he focused on satisfaction and benefits rather than deficiency through highlighting what vegetarianism *adds* to one's life. In basic terms, doing good makes one feel good about oneself, and animal organizations often mention this mental benefit in conjunction with moral messages about how their veganism prevents animal cruelty or environmental devastation.

In the previous chapter on values-based appeals, the sections "making a difference" (considered an altruistic value) and experiencing "moral integrity" (considered a personal value) exemplify how activists appeal to the mental health benefits of ethical vegetarian choices.

---

that for every wild-caught sea animal we see on anyone's plate, there are numerous other unseen animals who also lost their lives (Pew Environmental Group 2007).

In contrast, an emphasis on going vegetarian for physical health reasons, as opposed to mental/moral reasons, does not present as much opportunity to involve altruism and is essentially based on self-interested values. That being said, I don't mean to imply that self-*interested* is self-*ish*, as ensuring one's diet is healthy is a natural concern for anyone and is part of protecting one's own wellbeing (something every animal does).

My Concluding Analysis: In Support of Promoting Animal-Centered Altruism

In summary, considering all the messages' relationship to animal rights ideology, I was encouraged to find that most of the organizations prioritized altruistic values toward nonhumans over anthropocentric, self-interested values. Vegan Outreach and Compassion Over Killing put the largest proportion of emphasis on altruism toward nonhumans. Farm Sanctuary and PETA did as well, but, being larger groups, they also branched out into more anthropocentric issues (both altruistic and self-interested). The world needs people who are less selfish and less materialistic if we are going to end the mass exploitation of animals and nature. Communication messages that encourage altruistic or self-transcendent values in society can help change more than just a specific behavior; they help change worldviews and people's sense of self, hopefully motivating people to support causes on behalf of a wide variety of living beings (Crompton and Kasser 2009). When applied to vegan campaigns, this appeal to deeper, intrinsic values is not only an example of using the right means to an end, but it can also help animal activists reach one of their desired ends, as Maurer (2002) and the Humane Research Council (2007) note that emphasizing the immorality of meat-eating works to increase the number of people willing to go vegetarian (rather than just reduce some animal consumption).

In support of animal activists' creating a more caring society and demonstrating care for their own cause, I advocate that their messages should list altruistic appeals toward nonhumans first in all communication pieces that include a variety of rationales for veganism. Then they could mention other altruistic values, such as environmentalism and world hunger, followed by personal self-interest, such as health, at the end or in the smallest proportion. This prioritizing of the interests of *nonhuman* animals encourages people to overcome their humanist prejudices that view the interests of other animals as less important. Yet appeals to a human's own self-interest in health can also promote altruism if accompanied by the *mental* health benefits, such as peace of mind and pride, that vegans may receive from making a difference for others and acting with moral integrity. In creating pro-vegetarian campaigns, activists should see themselves not so much as marketers of vegan food *products* but rather as marketers of a social *cause*. And to more accurately reflect the animal rights organizations' primary commitment to the cause of nonhuman animal protection (and social justice more generally), their frames should prioritize altruism over human self-interest.

To further reflect this dedication to nonhuman animal protection in support of overall animal rights, activists should place greater emphasis in environmental frames on the negative effects of animal agriculture on wild/free animal species and their habitats. General discussions of pollution are open to interpretation to be perceived in terms of altruistic or self-interested concerns. This ambiguity can serve the utilitarian purpose of widening its appeal to a variety of readers who have different interests (Perelman and Olbrechts-Tyteca 1969), but it lacks the transformational purpose of explicitly trying to raise the level of respect people give to the interests of nonhuman species in comparison to human interests.

The next chapter supplements my empirical examination of activist materials by including my interviews with each of the organization's leaders so they can explain the rationales behind their framing choices. In this way, we not only examine the representation of animal rights rhetoric but its production as well.

Chapter 7

How Movement Leaders Explain Their Strategic Choices

Now that we have learned *what* these five animal rights organizations communicate in their vegan campaigns, it's useful to understand *why* they made those strategic choices. Through interviews in 2008, I gave the leaders of these organizations a chance to explain their decision-making rationales for crafting persuasive messages, particularly in terms of how they are influenced by their own views on nonhuman animals, humanity, and ethics. In this chapter I share some of the interesting commentary from each of the five animal rights organization leaders:

- Erica Meier of Compassion Over Killing (COK),
- Alex Hershaft of the Farm Animal Rights Movement (FARM),
- Gene Baur of Farm Sanctuary,
- Bruce Friedrich of People for the Ethical Treatment of Animals (PETA), and
- Matt Ball of Vegan Outreach.[1]

---

[1] I conducted these interviews by phone in 2008 and recorded them with each individual's permission. I have organized this chapter according to topics where I synthesize commentary from all the leaders as it fits the topic rather than providing full transcripts of each interview verbatim. I often paraphrase their comments, but I put verbatim comments/terminology in quotation marks. Note that Ingrid Newkirk, the co-founder and head of PETA, was out of the country at the time of my original interviews, so she recommended I instead speak with PETA's then-Vice President Bruce Friedrich. As VP of International Campaigns, Bruce was a fitting replacement since he had worked at PETA for over a decade at that time, heading up many food cam-

I asked them questions such as:

- To what extent and in what ways does your animal rights philosophy influence your message strategy?
- Explain the history of your food campaign message strategy and why you have chosen your current approach.
- In your campaigns, do you emphasize dietary changes based on the audience member's self-interested motives or more altruistic motives? To what extent does your choice affect how your audience members would or would not change their view of other animals?
- What values related to other animals do you assume the proposed audience member already possesses? What values related to other animals do you intend to promote in your food campaign message? Do any of these values conflict with each other? If so, how do you reconcile that conflict?
- In what ways, if any, do your campaign messages promote the similarities between humans and other animals? How is your strategic use of visual imagery related to how you would like your audience to view human beings in relation to other animals?
- Do you believe your campaign messages are influenced more by your theories on animal rights or your theories of what works best to get people to switch their diet? [2]

To structure their diverse responses to these questions, I organize their decision-making rationales into two communication ethics approaches – deontological (more idealistic and means-oriented) and utilitarian (more pragmatic and ends-oriented). I define deontological approaches as privileging the most ethical communication act (or right means) to achieve one's desired end result (as it is an ethical perspective that favors duty over consequences). Messages are assessed as

---

paigns. I was later able to interview Ingrid in 2012, and her comments are shared in the next chapter, along with recent feedback from all of the original five interviewees.

[2] It is interesting to note that, in answering my interview questions, many of the leaders voluntarily made reference to philosophical concepts and philosophers such as *Peter Singer, Tom Regan, utilitarianism, deontology,* and *pragmatism*. This speaks to the deep understanding that these activists have for moral philosophy, in particular animal rights ideology.

deontological if they are truthful/open, significant and important, sincere and representative of the speaker's identity, and/or reflective of the speaker's motivating values (in this case, animal rights ideology). I define utilitarian approaches as privileging whatever[3] communication means/message will result in the greater good (as it is a consequentialist theory ultimately evaluating actions according to whatever will maximize benefits and reduce suffering among affected claimants). In the case of vegan advocacy messages, utilitarian communication choices are based more on what will presumably work most effectively to create the greater good of helping nonhuman animals the most (getting people to reduce or eliminate animal consumption). For example, utilitarian appeals might pragmatically emphasize human self-interest more so than altruism toward animals, because they believe the former to be more motivating to the public, even though the latter appeal is more reflective of the activists' own motivations. Or utilitarians might stick with more mainstream/popular animal welfare appeals instead of making more challenging appeals to animal rights, as the former is more widely accepted and runs less risk of offending, even though the latter is more representative of the activists' own goals. These philosophical categories are useful to take into account, as they drive communication decisions at a fundamental level. And they relate to my thesis in favor of taking a more deontological or "ideologically authentic" approach of promoting altruism and animal rights ideology, yet I also hope that it will ultimately achieve the utilitarian goal of the greater good.

The two decision-making styles of deontological and utilitarian do not always contradict each other, as sometimes the most effective and expedient (utilitarian) message is also the most ideologically authentic (deontological). In this analysis, it is challenging to separate *animal* ethics from *communication* ethics, as they are not mutually exclusive categories and both include deontological (non-consequential) and utilitarian (consequential) aspects. My ultimate intention is to privilege ethical *communication* choices regarding how best to present animal ethics issues to the public.

---

3 There are some moral boundaries, as animal activists mentioned wanting to remain truthful.

WHAT THEY HAD TO SAY

My findings from the interviews reveal that, overall, most animal rights leaders use a blend of deontological and utilitarian communication approaches, especially in their choice to privilege nonhuman animal concerns over human concerns. But they often lean more toward utilitarianism in their choice to focus on animal welfare more so than rights, as the former has wider public appeal.

In the first section I discuss the leaders' justifications for choosing nonhuman animal-centered appeals versus anthropocentric/human-centered appeals. Then, I explore their decision-making rationales for factors such as which animal species they privilege, how they select visuals, and whether or not they choose to emphasize animal welfare or animal rights.

Justifications for Choosing Animal-Centered vs Anthropocentric Appeals

The most utilitarian/pragmatic organizations are Vegan Outreach and FARM, even though they each use different messages in pursuit of the same end result of encouraging people to reduce consumption of animal products. Vegan Outreach privileges animal-centered messages and FARM privileges human health and food-oriented messages. Matt Ball at Vegan Outreach believes that many people, especially college students, care about animal suffering. Vegan Outreach emphasizes animal cruelty in messages directed at this target audience because he believes this message is "strong"[4] and powerful enough to inspire some readers to actually change their diets. Ball thinks a self-interested health or environmental message may initially create more "agreement" about the need to change or greater willingness to "consider" the message, but it is ultimately not strong or compelling enough to spark such a significant change for most people. He explained a distinction in Vegan Outreach's communication goals, "We don't want people to listen. We want people to *change*."

To disprove the myth that self-interested appeals are more effective, he cited other health statistics, such as smoking or the obesity

[4] Note that I use quotation marks frequently throughout this chapter merely to indicate the interviewee's verbatim word choices, adding authenticity, and not to indicate the term's ambiguity or to challenge the term's social meaning.

epidemic, to argue that people will continue to consume or do unhealthy things even when they know it may harm them. He also feels the public will not believe an animal protection group's claims that all animal flesh is unhealthy over the current mainstream medical community's claims that chicken and fish are generally okay to eat. In addition, Ball stated, "a plurality of people who are vegan cite *ethical* reasons, animal issues, for being vegan," and they tend to be more committed to the diet, while many people who claim to be vegetarian, but still eat some chicken and fish, cite *health* as their main motivation. Therefore, Vegan Outreach chooses to primarily appeal to altruism toward nonhuman animals instead of human health. This is done for utilitarian reasons, but can also be viewed as deontological for an animal rights organization, making it both pragmatic and authentic. Similarly Erica Meier at COK also noted the utilitarian long-term benefits of focusing on nonhuman animals more so than human health:

> I think people who do choose the health aspect of vegetarian or vegan eating are less likely to change their view of animals….the next health fad that comes along that involves eating an animal product if it's for the purpose of them improving their health, then chances are they would probably stop being a vegetarian or vegan or choose the next health fad if it involves eating animals. I think that is another reason why we focus on animal cruelty because that is our goal to get people to change their view of animals.

Both Meier and Ball indicated that a permanent commitment to veganism is more likely if someone is motivated by concern for animals rather than just personal health.

Of all leaders I interviewed, Alex Hershaft at FARM had the most utilitarian viewpoint, as well as being the most doubtful in terms of his beliefs about human motivations and how people feel about other animals. He believes that most Americans think of other animals as "resources," with the exception of cats and dogs in many cases. Therefore, Hershaft thinks concern for animals is less influential at getting most Americans to change their diets than higher priority reasons such as the "availability, taste, cost, and ease of preparation of meat and dairy alternatives," concern for their own health, and concern for the environment (listed in order of how he perceives their influence). Notice that his list of perceived factors motivating the public goes from most self-interested at the top to most altruistic at the bottom. Therefore, FARM's food campaigns emphasize self-interested rationales such as the "attractiveness of vegan foods" over more altruistic rationales because Hershaft believes that appeals to self-interest are more likely to work when asking consumers to make a change as "fundamental as what they consume three times a day." He clarified, however, that if he is simply asking for a donation, then an altruistic appeal on behalf of other animals may work better. But in requesting that consumers make a larger "sacrifice," such as "changing life-long habits," he does not think altruism is a strong enough appeal.

No other leader said his/her group primarily privileges appeals to human self-interest, as the rest prioritized animal cruelty messages, which is more of an ideological/deontological communication approach than FARM's, considering the main function and concern of all these organizations is nonhuman animal protection. Gene Baur at Farm Sanctuary did admit he thought the health argument was probably the most convincing at getting people to go vegetarian. Yet, despite this belief in the power of human self-interest, Farm Sanctuary still uses concern for farmed animals, not health, as their main message because Baur says Farm Sanctuary is trying to appeal to people's hearts. Therefore, his communication decisions are more authentic/deontological than pragmatic/utilitarian in this aspect. Even though the end result in both the content of Farm Sanctuary and Vegan Outreach is similar in their emphasis on farmed animals, Farm Sanctuary's choice is more influenced by deontological or ideological concerns while Vegan Outreach's is more utilitarian.

Utility and ideology can overlap. For example, Erica Meier of COK noted that asking someone to be altruistic enough to spare animal lives is in someone's personal self interest because it ultimately makes that person feel good about him/herself. She said:

> I would say that we are probably trying to appeal more to an altruistic sense but in that sense it might make somebody feel better, it might be more of a personal choice to make them feel better to move away from supporting that kind of suffering that they didn't realize they were supporting.

Bruce Friedrich at PETA also noted the self-interested benefits people will experience by being kind toward animals on a morally consistent basis:

> Mostly we focus on altruism… saying "you oppose cruelty to animals and if you are eating meat, then you're supporting cruelty to animals." Everybody knows that people want to be consistent, and eating meat is inconsistent with what people think about their own motivations and ethics.

These beliefs express what I discussed in the last chapter – how being altruistic towards others can also be in one's own best interest, as one experiences mental health benefits from acting upon one's values toward justice and compassion. Yet, the challenge for activists is to get past people's rationalizations of their contradictions as they cognitively seek to avoid the dissonance we are trying to activate.

Animal-Centered Messages

Choosing whether to emphasize animal welfare or animal rights

Gene Baur explained that his organization "marries" ideology and utility in the overall message strategy. For example, within the spectrum of animal altruism appeals, Farm Sanctuary, like most animal organizations, is more pragmatic in their communication decision to reach people "where they are." Baur believes most people already care about animal welfare, so he does not seek to move them to animal rights but rather asks them to simply "evolve" to expand their welfare concerns out from companion animals to farmed animals. This aspect of the communication decision is pragmatic because it is easier and probably more immediately effective not to challenge people's basic beliefs about animals to a great extent.

All the animal organizations are pragmatic by choosing not to challenge people's basic speciesist worldviews and simply appealing to people's existing, mainstream animal *welfare* concerns about suffering and cruelty. This strategic route is perceived as an easier way of gaining acceptance than it would be through appeals to animal *rights*. Baur said, "I don't think there's a conflict, really, between the values that we're promoting and the values that most people hold," which he described as "humane" and "compassionate." He said, "We hope to tap into that sentiment and encourage people to act in ways that are consistent with their values, and most people want to see themselves as compassionate." Similarly, Bruce Friedrich explained PETA's choice to tap into the public's existing desire not to be cruel to animals:

> Everybody I've encountered in my speaking with people about animal issues is opposed to cruelty, so it's really just a matter of helping them to understand that their own food choices, if they are eating meat, are out of integrity or out of compliance with their basic values.

Most leaders use the word *evolve* or *extend* to suggest that the main change in attitude that needed to happen was for Americans to transfer their existing animal welfare concerns about companion animals over to animals who are used for food. While animal protection leaders do think that most Americans believe farmed animals (except for fish) feel pain and do have some basic emotions, leaders assume that most people simply have not considered farmed animals as sentient individuals in the same way they consider dogs and cats sentient.

The American public is not typically asked to think about farmed animals and factory farming, as they are purposely kept out of sight and out of mind. Therefore, these animal organizations make it their job to get the public to "open their eyes" (a popular phrase used in the interviews) and consider farmed animals and the role consumers play in animal suffering. Baur said, "We want people to question the status quo – to question if what we are doing to animals is appropriate." He said that Farm Sanctuary is challenged to "encourage people to be somewhat introspective," and "that's a hard thing to do because people have to do that on their own and we want to provide the, sort of, the nudge that gets people looking internally and looking honestly at their own behavior." This speaks to the challenges of creating persuasive appeals that people will want to listen to while also containing a critique people may not want to hear.

In crafting their persuasive appeals, most animal groups, with the exception of FARM to some degree, use a two-pronged message strategy designed both to:

1) Raise public awareness about farmed animals as sentient beings in comparison to other familiar nonhumans like dogs and cats, and
2) Inform people that farmed animals are suffering greatly, especially in factory farms.

For example, Baur explained the need to show the public that farmed animals both can suffer and do suffer:

> In order to protect something or somebody, it is very helpful to know and understand that something or somebody. A big part of Farm Sanctuary's message is that these are living, feeling beings and they suffer just like your cats or your dogs might suffer. So we try to make it *relevant* for people that way. That they are currently being abused in mass and it's just, just, just … wrong.

By using analogies to make farmed animal suffering relevant, in the way Baur presumes our own dog's or cat's suffering is relevant to us, he hopes people will begin to see farming as abusive. Similarly, Meier explained the two-pronged goal like this:

> I think that our basic goal in terms of animal liberation related to farm animal issues in our specific campaigns is to encourage people to see farm animals differently… And so the first breakdown is to get people to see these animals for who[5] they are. They are no different than the dogs and cats we bring in to our homes. They are no different than the wild animals who we see at a distance and have a great affinity for. They experience emotions. They feel joy, and pain, and sorrow. So that's one issue – sort of breaking that down. The other issue is getting people to understand that these animals are not leading the happy lives that they think they are. These animals are suffering tremendously behind the closed doors of factory farms and slaughterhouses. And so to get people to see the reality of how these animals are being treated is one of our main goals.

Notice Meier's description of the first goal of seeing farmed animals *differently* means seeing them *similarly* to the way we see other animals who we already care about. And her second goal is to get people to see them actually suffering in modern farming. Her repeated use of the word "see" lends power to the visual.

Choosing visuals of nonhuman animals

Most animal organizations follow the aforementioned strategy of persuading people of farmed animals' individuality and simultaneously informing them how much these individuals suffer in agriculture. Often leaders say these two goals are accomplished with a parallel combination of two types of visuals:

1) Showing *happy* photos of contented animals displaying their "personality" (per Baur) and just being "who they are" (per Meier), and

---

[5] Note her intentional use of the descriptor "who" that helps to personify the farmed animals as *someone* in a way that inanimate descriptors like "that" or "it" would not.

2) Showing *sad* or graphic photos of animals suffering in factory farms, stockyards, and slaughterhouses.

For the former "non-abuse" (happy) visuals, Meier, Baur, and Friedrich all said that they chose photos that allowed viewers to look into the faces, and particularly the eyes, of the farmed animal. Baur said, "animals' eyes, like humans' eyes, can often times say a lot, and looking into the eyes can provide a real connection." In an attempt to create a connection in viewers' minds between the similar sentience of farmed animals and companion animals, Matt Ball explained that Vegan Outreach likes to use photos that show people interacting with or petting farm animals, so "people can see pictures of people in a way that looks like a person with their cat or a person with their dog, but it's with an animal that they are generally used to eating." Visuals can create connection both by allowing us to look into someone's eyes and by allowing us to visualize ourselves befriending them.

Toward the second (sad) type of visual, when choosing factory farm photos, Baur said he wants the photos to "touch people viscerally" and "to, I don't want to say shock but, to expose the realities of factory farming." Friedrich noted that PETA wants those shocking factory farm photos to create "empathy." To generate empathy, Meier explained that COK uses confinement photos (ex: hens in battery cages) more than slaughter photos because she thinks the public can better "relate" to being confined than they can to being slaughtered. She hopes these confinement images might facilitate people "putting themselves in that situation or seeing their dog or cat" in a crate. Ball said Vegan Outreach makes utilitarian decisions about using factory farm photos that are "powerful" enough to be emotionally affecting without being so "gory" that people will avoid reading the booklet, especially cover photos:

> We don't want to pick the goriest pictures to give people more of an excuse to write it off as propaganda, but we don't want to tone down our message so much that even the people that say "I can't look at that" will look at it because it takes away too much of the power of the message – the reality of what goes on in factory farms. We are trying to be somewhere in the middle that will influence the most number of people.

He explained that this idea of settling "in the middle" of the abuse spectrum is done for the utilitarian purpose of creating "the most change per dollar spent in an hour spent leafleting." It speaks to the need to balance appeal with critique.

While all the leaders believe they are showing the public the "reality" and "truth" of what goes on in factory farms, Friedrich, Hershaft, and Ball specifically mentioned visual honesty in the interview. Friedrich said PETA's images are "a *representative* sample of the abuses that are standard." Hershaft said FARM used, "whatever works. I mean as long as it doesn't distort the *truth*. We focus on whatever we feel would catch us more attention." So, FARM's ends-oriented rationales are tempered by a means-oriented concern that communication be truthful. Ball emphasized truth more than drama for Vegan Outreach:

> We're trying to use pictures that *honestly* represent what goes on in factory farms ... We don't want to go for gore for gore's sake. We want to have pictures that are *defensible* in terms that this is the *reality* of what goes on – this is standard practice – and not have people think that it is sensationalized propaganda.

In general, *truth* was an important concept mentioned in the interviews, not just in terms of the organizations being truthful in their own communications but also in terms of sharing the ugly truth that agribusiness hides from the public.

Choosing which species to emphasize

Ethical choices are also reflected in organizations' decisions of which species to highlight. For example, most animal protection organizations avoid talking about fish as sentient beings, although the food materials usually address sea animals to some extent, such as making fishing and aquaculture an environmental issue. This choice to use an environmental (rather than sentience) frame for fish is often done for utilitarian reasons, as there is not mainstream public acceptance that fish even feel pain, much less have personalities. PETA is the only group who is willing to tackle this challenge in actively promoting fish sentience, including personalities, making PETA the most ideal-driven/deontological group on this issue. Yet, all animal organization leaders believe fish are sentient and know they are killed for food in greater numbers than land animals, so from a means-oriented/deontological communication standpoint fish should be prioritized. Meier laments that as a small group, COK does not have the resources to emphasize fish as much as they would like and admires PETA's efforts. And while I did not ask Baur specifically about fish, it seems reasonable that Farm Sanctuary is somewhat excused because their sanctuary rescues only land animals, so aquatic animals are not their priority.

Ball admitted that Vegan Outreach makes an intentional compromise on the fish issue because even though the group's goal is to reduce animal suffering as much as possible, he thinks people will dismiss their message entirely if they see an animal welfare appeal toward fish, as that requires a larger attitude or values change than most people are willing to make. So, Vegan Outreach has come to largely avoid fish messages for utilitarian communication reasons because it may compromise their success at getting people to stop eating land animals. And in keeping with their utilitarian communication and animal ethics goals, Vegan Outreach has begun to prioritize the welfare of factory farmed birds and pigs, as Ball believes they suffer in the greatest proportion and numbers of all land animals. Therefore, Vegan Outreach's materials feature many photos of birds and pigs and ask people especially not to eat their eggs or meat.

Ball explained that the decision to emphasize birds is still a compromise in possible effectiveness because the public tends to sympathize more with mammals, such as cows and pigs, but he also acknowledged birds still rank higher than fish in public sympathies. He explained that because birds make up the vast majority of land animals killed, in part because they are smaller than mammals, Vegan Outreach does not want to prioritize mammals just to gain greater reader acceptance while risking increasing the trend of people giving up red meat and switching to poultry. Meier expressed similar concerns and COK also prioritizes birds for these reasons.

In some ways, the deontological/ideal-driven focus on birds can be considered means-oriented and the pragmatic marginalization of fish can be seen as ends-oriented. Animal organizations sacrifice wider public acceptance of their message in favor of attempting to save the largest number of animals from suffering, birds. This is a balancing act and risk that COK and Vegan Outreach are willing to take on behalf of birds, but are less willing to take on behalf of fish, despite their equally vast or greater suffering, because of pragmatic concerns that the risk would not pay off with sea animals. However, COK does include a few fish messages, suggesting they are not as concerned as

Vegan Outreach that mentioning fish will turn people off to the whole idea of ethical vegetarianism. In utilitarian fashion, Hershaft said FARM more frequently uses pictures of pigs and cows than birds. These framing choices to highlight more appealing species (or those who more closely resemble humans) exemplify the animal rights movement's challenge between operating around speciesist prejudices versus directly confronting these prejudices. Another prejudice to confront is human's status as an animal.

Choosing whether to compare human and nonhuman animals

All of the leaders believe there is a similarity between humans and other animals, as they agree we are all animals, but when I asked each of them whether promoting similarity between humans and other animals was part of their strategy, only Baur and Friedrich said it was. Gene stated, "Inherent to our message and to our mission is the recognition that the other animals have feelings and value and interest in their own right." And Friedrich suggested, "For the same reason you wouldn't eat a human being, you shouldn't be eating a dog or a cat or a pig or a fish." And he added:

> In our farmed animal campaigns we keep the focus on the fact that other animals feel pain in the same way and to the same degree that humans do, and consequently there is not a moral difference between inflicting pain on a human being or inflicting pain on anyone else who experiences that pain to the same degree, which is of course at the very least mammals, birds and fish.

Alex said FARM only "tangentially" promotes similarity between humans and nonhumans in their life-affirming messages conveying that we all have the right to have our basic needs met, and "the most fundamental need is the need to live." Because promoting similarity between humans and nonhumans challenges the accepted human/animal dualism and seeks greater changes in attitudes from the public, I see it as an ideologically-authentic, deontological approach.

While humans may occasionally be compared to other animals, none of the animal rights organizations actively promote or privilege a similarity between humans and other animals in the advocacy materials I studied. Ball said Vegan Outreach doesn't compare humans and nonhumans "not in terms of the physiological or evolutionary continuity with humans," but they do appeal to people's emotional connection with nonhuman animals as companions. Rather than comparing humans to nonhumans, Meier explained that it is less challenging for COK to show people that different types of *nonhuman* animals are similar to each other:

> I think the average person in our society has a very distinct line between humans and animals. And by appealing to their emotions for animals – because people obviously have compassion for animals – by appealing to that emotion that is definitely there we are hoping to sort of erase that arbitrary line that our society has drawn between cats and dogs and farm animals.

Confronting people's humanist self-perception that they are morally distinct from the animal kingdom is a risk that, for utilitarian reasons, animal rights activists do not often seem to take at this stage in the movement.

Choosing whether to include welfare reform messages

Friedrich and Baur explained why they include an animal welfare "reform" message (reforming industry or laws), and Meier and Hershaft explained why they did not. Meier expressed concern that consumers want to "justify eating meat" by choosing free-range products or generally assuming the animals on all farms are relatively "happy." To explain the industry's role in facilitating this "misunderstanding" so that consumers think animals are treated humanely, she noted:

> It's something that the industry is recognizing as extremely lucrative, so a lot of the messages that the industry are starting to use like the "happy cow" milk campaign, and in the egg industry the "animal care certified," and Oscar Meyer has kids singing songs about how great it is to be a hotdog. I think that the industry is recognizing that more people in our society are seeing the truth about factory farming, so they are trying to appeal to their emotional side as well by saying "Oh it's okay. These animals are all happy." And a lot of people are buying into that.

Hershaft agreed with this sentiment, and thinks farm animal welfare reforms by animal rights organizations are counterproductive because they may work to ease consumer guilt enough for people to continue eating animals. He said of fellow animal activists:

> When they advocate bigger cages and an occasional ray of sunshine for these animals as they continue being raised for food, they are providing the medicine – the band aid, the aspirin – that the socially-conscious consumers are desperate for in order to keep perpetuating the problem of eating animals.

Hershaft referred to public opinion statistics about how Americans overwhelmingly say they don't want farmed animals to suffer, yet they still eat them. He felt these contradictory attitudes were indicative of the role that humane reforms play:

> American consumers want to continue eating animals, and at the same time have a clear conscience that they are not being mistreated. It's a perfect solution for the consumers. It's a perfect solution for the meat and dairy industry. It's a win-win situation. The only losers are the animals. And of course as animal rights advocates, we cannot stand for that.

Focusing on the means, Hershaft stated that animal organizations should be sincere and consistent in promoting their belief that animals should not be used for food. He believes reforms are "counterproductive" coming from animal rights groups because it:

> Gives the impression that we approve of the use of animals – exploitation of animals – for food as long as they are treated a little less reprehensibly. We feel that welfare reform is something that the animal exploiting industry should be introducing to try and entice the consumer, the socially-conscious consumers, to consume them.

Baur agreed that animal rights organization messages should make it clear that industry reforms do not make industry "good" and that reforms are not better than veganism. But he still argued that a mix of welfare messages (enact legal reform) and rights messages (go vegan) can work at the level of *strategy*, even if he admitted they are somewhat contradictory at a philosophical level. He said that, philosophically, Farm Sanctuary is an animal rights organization, but "from a broader strategic standpoint and broader messaging standpoint, the movement exists more on a continuum. Not one block of rights people and one block of welfare people." This belief influences strategy, as he thinks, "Welfare folks often times gravitate toward and evolve toward rights folks and a rights position over time," so Farm Sanctuary's legal reform messages are a "practical near-term approach," while their vegan messages work on a "broader societal cultural shift that has to happen that goes beyond laws." This indicates a reformist approach in the legal realm and an abolitionist approach in the realm of consumer culture.

Baur sees humane farming reforms pragmatically as "incremental abolition," as "steps in the right direction" meant "to abolish certain cruelties in a hope that we eventually create the humane vegan world we dream of." In reform messages Farm Sanctuary capitalizes on popular welfare sentiments, arguing that factory farming practices are "cruel" and "outside the bounds of acceptable conduct in a society that values compassion." To explain his organization's approach to both welfare and rights messages, he said:

> The welfare reforms are often times seen as soft within the animal movement. When it comes to welfare stuff our messaging is hard. Ban the crates. Ban this. Ban that. But when it comes to the rights, which within our movement has tended to be more strident, we put a little soft edge on that and encourage people to adopt a vegan lifestyle. So that is kind of how we have taken those two aspects of our movement to try to kind of marry them.

In this way, the legal reform messages are framed more critically, to give them an edge, while the abolitionist vegan consumer messages are framed more gently, to take the edge off.

Friedrich uses both deontological and utilitarian logic to articulate why it makes sense for animal rights organizations to promote "less abusive production" methods in farming. He begins here with a focus on how welfare reform is the right means to an end by referencing deontological philosopher Tom Regan:

> Both from a pure animal rights-Tom Regan-perspective, if you say, "How would I want to be treated if I were that animal?" obviously you want to have the worst abuses eliminated. And then, of course, from a utilitarian standpoint, it seems to move us further toward a world that we are envisioning to treat animals not as badly.

He indicated that reforms are better for the animals both now and later, as reforms may encourage incremental abolition over time. To explain why the last sentence of his quote is utilitarian, Friedrich argued that there are higher rates of vegetarianism in countries where "there's more consciousness and more 'humane' production," as humane laws help raise people's awareness about farm animal suffering, so more people might then withdraw their support. He stated:

> We do have to get to a point where people say, "Yes, chickens shouldn't be caged. Yes, pigs shouldn't be crammed into crates" in order for more people to make a choice not to eat animal corpses at all.

However, previous comments from he and other leaders suggest that Americans already believe intensive confinement is cruel and, therefore, wrong. So, I suggest that the possibility of getting rid of the worst cruelties will not necessarily *change* consumer attitudes or behaviors. Perhaps this is why Friedrich claims that regardless of consumer attitudes, legal reforms mainly are better for *existing* animals who currently must endure the conditions of industrial agriculture. This latter point attempts to frame welfare reform as an ideologically-authentic means to an animal rights end by emphasizing how it treats fellow animals as we humans would want to be treated, if we were in their unfortunate circumstance. But Friedrich's other point is ends-oriented in arguing that welfare laws will result in increased awareness and public scrutiny of farmed animal cruelty, which may lead to increased vegetarianism.

Simplifying animal rights ideology

Animal rights organizations in my study, especially COK, Farm Sanctuary, and PETA, often simplify their deeper animal rights philosophies to create communication strategies based on shallower, or more popular and broad-based, ethical sentiments. Therefore, a relationship exists between theory and strategy, but it is just partial or more at a surface level. Baur commented how he thinks messages do not have to be either animal rights-based or animal welfare-based. Rather, for him, these messages exist on a "continuum." Despite being a rights organization, Farm Sanctuary's messages tend to avoid using the word *rights* in favor of the word *compassion*, which serves to bridge animal rights and welfare. Baur stated:

> The word "compassion," I think, is very important in the animal movement… in my view, it is a strong word and it embodies what our movement should be about. And it's not divisive within our movement like rights versus welfare has become.

For utilitarian reasons, it is more effective to avoid any controversial or ideological terms in favor of a more neutral, but equally representative, term such as "compassion." Similarly, Friedrich explained that

PETA prefers to focus on promoting compassion and to "stay out of the more academic utilitarian versus deontological versus whatever you want to call Gary Francione's ethic discussion."[6] While Friedrich thinks all animal rights groups agree with the basic premise of PETA's philosophy that "animals are not ours" to use as resources, the focus of animal organizations' work and messages should be on ending the actual *use* of animals. He said, "PETA tries very hard to focus on 'brass tacks' issues rather than to get mired in any sort of philosophical disputes." This fits with his belief that the public also agrees that much of the animal cruelty, in practice, is wrong, so that we don't have to work to change their philosophies. In favor of utilitarianism, Friedrich said PETA will work with any group who is "trying to make the world a kinder place," even if they do not share the same ethical philosophies in all aspects.

COK also simplifies philosophy in their messages by making the whole popular idea of being compassionate toward animals "simple" for the public to practice toward farmed animals, as the organization's messages highlight the *ease* of ethical vegetarianism. Meier said COK's goal is to:

> make the idea of not eating animals a mainstream issue – to bring it to the forefront, make it a household term, make it *accessible* to people, make them realize how *easy* it is to simply stop eating animals.

In order to help make veganism mainstream, she said COK has gravitated toward providing more practical guidance on *how* to be vegetarian and not just ethical rationales on *why*. "We are now trying to offer the general public a pragmatic view of *how* they can take steps to help animals," Meier explained. "We try and offer *tools,* not just providing them with reasons why they should be vegetarian or vegan." Like PETA, COK's priority is changing harmful *behavior* toward nonhuman animals more so than changing harmful beliefs toward nonhuman animals.

---

[6] Animal rights legal scholar Gary Francione takes an abolitionist view in favor of animal rights over welfare reforms, although some reforms, such as banning foie gras, can also be abolitionist. See my discussion of his philosophies in chapters two and three.

Separating animal rights ideology from strategic communication

Rather than simplifying ethical philosophy, some animal rights organizations attempt to separate their ethical philosophies from their communication strategies, instead, focusing on what works best. The latter is especially true for FARM's and Vegan Outreach's utilitarian strategies. Ball explained that the differing animal ethics philosophies of he and his Vegan Outreach co-founder do not have to be perfectly in sync in all aspects. Vegan Outreach focuses on where these views "converge" with the views of the public on the issue of suffering. This is similar to Baur's and Friedrich's belief that there is much public consensus that cruelty and suffering is wrong. Ball stated:

> We don't have to come to an agreement of what animal rights or animal liberation is between us because the bottom line is that there is so much suffering that it doesn't really matter if you're a deontologist or a utilitarian.

Ball explained that Vegan Outreach does not present information to the public "in terms of animal rights" nor is it "based on philosophy." "We're not trying to have people agree with Tom Regan or Peter Singer," he explained. "We're trying to reduce the amount of suffering as much as possible." In support of his utilitarian emphasis, Ball's message content is "based on what we've found over the years that has been effective at creating the most amount of change in people's habits." Essentially, Vegan Outreach does not feel the need to persuade people about ethics but rather just provide them with consumer "information" about factory farming that will likely offend the moral beliefs people already hold, as most people have an "inherent rejection of cruelty." He explained, "Our message is more a matter of *presenting information* – the reality of factory farms – to people so that they can see these things and make an *informed choice.*" In this way, I would say Vegan Outreach takes the role of an investigative journalist or documentarian in exposing wrong-doing rather than proactively attempting to persuade people to reconsider their beliefs about right and wrong.

FARM also seeks to meet the public where they are, rather than trying to get them on the same page ideologically with animal activists. Therefore the strategic rationale that FARM uses in their messages differs from their organization's ideological rationale. Hershaft expounded:

> Well, our message strategy is always determined by our audience...But our motives are to – those of bringing justice and fairness to the rest of the animal kingdom that we have been exploiting so ruthlessly.

He admitted that strategy and ideology are separate with the statement, "If we appeal to their self-centered interests by talking about the desirability of vegan foods, it has nothing to do with their view of animals." I also see this as an admission of the limitation of this food-oriented strategy in helping animal rights overall. Even though Hershaft believes the value of animal rights movement is that it improves human sensitivities, he still does not advocate for dedicating much time to emphasizing an empathetic message or trying to promote animal rights because this requires more resources than his group has. He explained, "We feel that that's too difficult an issue for a small organization to tackle. So we really don't try to change American values vis-à-vis animals." Here there is a utilitarian emphasis on what improvements are realistic to achieve rather than on ensuring those improvements are made for the right reasons.

Hershaft also clarified that even FARM's one campaign that emphasizes altruism toward nonhuman animals, World Farm Animals Day, is more informative about welfare issues than it is transformative about rights. The slogan of the campaign is "Lest we forget their suffering" and is dedicated to mourning and memorializing the billions of animals who die in slaughterhouses and farms. He said that campaign only affects the public's "views of the *treatment* of animals. It's not their views of animals themselves," meaning it mainly changes views toward animal *agribusiness*. He even dismissed the strategic value of the campaign by saying, "We don't feel that World Farm Animals Day really does as much to advance our goals as some of our other campaigns, but we just do it out of a sense of obligation." This latter deontological statement clarifies that World Farm Animals Day is ideologically motivated, but the fact that Hershaft perceives it as less effective reflects his overall utilitarian orientation.

When it comes to animal rights, Hershaft said the movement is generally more about humans than it is nonhumans, as it ultimately encourages society to be more sensitive and caring. This shows how animal rights can be both selfless and self-centered. He said he came to this realization based on the slogan of an old animal rights documentary, *The Animals Film*:

> Their slogan was "animal rights – it's not about them. It's about us." And I was really intrigued by that slogan. You know people ask me "why do you worry about a mosquito? Why do you worry about a mouse?" And I explain to them "It's not about the mouse. It's not about the mosquito. It's about me and my attitude about life – towards other living beings." And this is the, really to me, the ultimate value of the animal rights movement is to make people more sensitive to the suffering of others. It's not about any particular animal. It's about us!

Hershaft's focus here on the movement's goal of improving the character of our own species indicates a belief in the value of virtue ethics – an ethical theory advocating that individual moral development will lead to moral behaviors. While in one sense the notion of improving human moral integrity is the thesis of this book, I also believe that we do this for other animals, not just for us. I wouldn't want us to revert to the self-centered or anthropocentric belief that kindness toward nonhuman animals is worth cultivating primarily because it ensures kindness toward fellow humans. Nonhuman animals are inherently valuable individuals, worthy of our respect, regardless of whether our society/species receives any personal benefits from the moral development that this recognition naturally produces. I agree with Hershaft's statement that animal rights is largely about us humans, but I would add that animal rights is not just about us becoming a more responsible and caring species but also about us starting to see ourselves in a humbler light as a fellow animal sharing the earth.

MY ASSESSMENT OF DECISION-MAKING CHOICES IN STRATEGIC COMMUNICATION

The fact that I organized the leaders' interview commentary into the moral philosophy categories of deontological (means-oriented) versus utilitarian (ends-oriented) indicates my belief that moral philosophies are fundamental to influencing decisions about how to craft persuasive communication (and evaluating their soundness). I personally favor a deontological/ideologically-driven basis for making ethical communication decisions as this encompasses more integrity and less moral relativism.[7] My own preference for deontological ethics in communi-

[7] This does not preclude a need to use utilitarian decision-making either as a supplement or as a back-up to deontological decisions, as utilitarianism can add necessary flexibility as well as potentially lead to the noble goal of creating the greatest good for the greatest number. I also recognize that sometimes utilitarian and deontological

cation, and largely in animal rights philosophy too (favoring inherent value of individual life more so than reduction of suffering), influences my evaluation of the strategic value of the activist leaders' rationales here. My goal is to offer constructive insights (to their own) regarding how best to strategically design communication messages to achieve animal rights goals.

Based on interviews I did with the five organization leaders, I found that most leaders used a blend of deontological and utilitarian approaches to communication decision-making. All organizations except FARM were more deontological in their overall choice to be more animal-centric than anthropocentric, but within the spectrum of these animal-centric appeals, they all often leaned more toward utilitarianism in their choice to privilege animal *welfare* frames over animal rights frames, as the former is more widely accepted. Activists asked people to "extend" their existing animal welfare concerns out from companion animals to farmed animals.

Leaders tended to use a two-pronged strategy of first getting people to recognize farmed animals as sentient beings and, second, to recognize how much farmed animals are suffering in agribusiness. The first tactic is more aligned with an animal rights perspective to see farmed animals as fellow subjects, but the transformational aspect of the frame could be improved by including more comparisons to the *human* animal. The second tactic is more aligned with a welfare perspective to emphasize a cruelty and suffering frame instead of putting these within the context of an overall injustice frame critiquing the ownership, breeding, and exploitation of the bodies of fellow subjects.

In deciding which visuals to use to highlight the animal cruelty frame, all animal organizations were deontological in choosing images that are truthful and reflective of standard agribusiness practices, but they were utilitarian in seeking pictures that would emotionally affect people enough to hopefully inspire change. To add to this utilitarian perspective, Matt Ball also stated these selected emotional images could not be so disgusting that they kept people from looking.

In selecting which species to feature in visuals, PETA was the most deontological by including the animals, fish, with whom Americans least identified. Vegan Outreach was the most utilitarian in largely

---

philosophies both end up advocating for the same means to an end, albeit possibly for different reasons.

excluding fish, but all other organizations also marginalized fish in favor of land animals, despite the massive numbers of sea animals who are used for food. FARM and Farm Sanctuary could be said to privilege mammals the most, as they believed Americans most identify with fellow mammals, so Vegan Outreach and Compassion Over Killing were deontological in their choice to privilege birds, as birds are the land animals who are most exploited. I believe it is most ideologically authentic to show animal species in relative proportion to the extent to which they are used for food, which would place sea animals and birds as the species most in need of attention. We likely won't be able overcome the massive discrimination that non-mammals face if we in the animal rights movement do not cultivate a belief that non-mammals matter too.

Regarding the connection between animal rights philosophy and the organization's message strategy, leaders either separated the two or simplified deeper philosophies to gain greater consensus at a shallower level. FARM, and Vegan Outreach to an extent, tended to take the former route of selecting strategies largely based on utilitarian concerns for promoting whatever rationale works best to get people to go vegan or to reduce consumption of animal products rather than privileging messages that best promote animal rights ideology. On the other hand, COK, Farm Sanctuary, and PETA were more ideologically-driven but still ultimately utilitarian in their preference for simplifying animal rights ideology in their messages so that messages appealed to more widely-accepted aspects of animal ethics, such as compassion and welfare. Vegan Outreach did this too, but Ball was overtly utilitarian in his admissions that (1) he knew Vegan Outreach's focused behavioral messages were limited to helping only *farmed* animals and not animals in other exploitative situations, and (2) he might be willing to discredit animal rights if it would get more people to stop eating animals. This illustrates how an animal rights organization can emphasize animal cruelty, an ethical issue, in support of an animal rights goal, veganism, yet not be committed to promoting the kind of ethical worldview that is authentic to animal rights ideology and serves the broader, long-term goals of the animal rights movement. My concern is that this constrains the discourse to welfare issues for farmed animals instead of transforming the discourse to encourage people to reevaluate their relationship to animals in all situations.

Regarding the animal organizations' tendency to embrace popular values and consensus, many leaders did not see themselves as producing persuasive messages. These leaders seem to see their message strategy as more *informative* than persuasive, as they often described their messages as providing "facts" and "reality" and appealing to animal welfare values that the public already holds. I concluded that these animal activist communication messages could be conceptualized as a more progressive version of journalism that is largely providing a different and untold perspective on agriculture so as to enable consumers to make more informed choices in the marketplace – choices that will fit with their personal values and priorities. While the public needs to be educated and learn the untold story of the problems associated with an animal-based diet, and while it makes sense for animal organizations to appeal to commonly-held values for cultural resonance, animal activists should not shy away from embracing a more openly persuasive role in emphasizing key altruistic values that they extend, amplify, or transform to support an animal rights philosophy.

While being more informative than overtly persuasive may seem more democratic or even more ethical, it is not any less ethical for an animal rights organization to serve in their role as an advocacy organization that is trying to argue for a change in attitudes based on the moral principles that the organization holds. To be more openly persuasive, as long as the advocacy organization is not misleading the public or concealing its advocacy status, is more honest than trying to provide more objective information that aligns with the public's values more so than their own (Freeman 2009b).

The remaining two chapters comprise the third part of the book, primarily encompassing prescriptive guidelines on how to best construct animal rights messages. In the next chapter, I continue my discussion with activist leaders about their most current farmed animal campaigns, advice for best practices, and predictions for the future of animal advocacy.

Part III:
Strategic Communication Recommendations for Vegan Activism

Chapter 8

Activists' Latest Insights and Projections

In this chapter I share insights and updates from my more recent interviews[1] (July 2012) with the five leaders whose organizations I studied: Matt Ball of Vegan Outreach, Gene Baur of Farm Sanctuary, Bruce Friedrich of Farm Sanctuary, Alex Hershaft of the Farm Animal Rights Movement (FARM), and Erica Meier of Compassion Over Killing (COK). Plus this time I also interviewed Ingrid Newkirk of People for the Treatment of Animals (PETA). The addition of Ingrid was important for several reasons. Not only is she the founder and director of the largest animal rights group in the world, but I needed to ensure that I had someone from PETA represented now that Bruce Friedrich has moved from PETA to Farm Sanctuary.

I'll begin by sharing their latest campaigns, as part of their overall strategy to address core cultural issues that cause exploitation. This includes a discussion on which values they believe to be most important to emphasize. While they obviously believe humans are animals, they explain some cultural concerns about overtly making the human-animal comparison. They'll discuss their opinions on appeals to reduce meat consumption in relation to appeals for veganism, as well as how to challenge *all* types of animal farming, not only factory farming. I'll end with their projections for the future of animal activism, both in terms of what concerns them and what gives them cause for hope.

---

[1] I conducted these interviews by phone. In this chapter I use quotation marks to indicate the activists' own direct words/phrasing. Otherwise, I paraphrase them. The activists speak for themselves and not necessarily on behalf of everyone at their organizations.

LATEST & GREATEST VEG CAMPAIGNS

When I asked leaders to describe their newest or most exciting campaigns, they were eager to share them. I notice that all the campaigns related in some way to making it easier for people to be vegan, so that consumers feel supported in their choice. For example, Alex Hershaft of FARM is excited about their new strategy of "integrated vegan advocacy" that he describes as "revolutionary." This integrated approach nurtures new vegan recruits for at least six months so they will be more likely to stick to their pledge of eating more vegan meals. The pledge is part of FARM's new 10 Billion Lives tour where a large van equipped with eight video screens and 32 headsets goes from city to city, often following rock concert tours. At the van, activists pay passersby a dollar to watch a four-minute factory farming video. After this "shock element" at least 80% of viewers are then more open and receptive to a request to sign a vegan pledge, where they select the number of days a week they plan to now eat all vegan meals. While shocking video and a vegan pledge may not be totally new tactics, what is different is the addition of FARM activists staying in touch with the pledgees for the next six months to support them with more information, remind them of their commitment, celebrate how many animals they are currently saving, and encourage them to extend that life-saving pledge to seven days. Hershaft hopes this reduces vegan recidivism rates, as witnessing upsetting videos may suspend people's "carnistic mindset" for a few days, but the emotional effect doesn't last in a carnistic culture and requires some prolonged encouragement.

In a similar attempt to address vegan recidivism, Gene Baur and Bruce Friedrich of Farm Sanctuary, noted the power of social support, something that is driving their Compassionate Communities campaign. These are grassroots campaigns in cities across the nation where Farm Sanctuary supports local vegan food movements, building networks and communities. This not only includes social support but also has a business component that encourages local restaurants and retailers to sell more vegan items and avoid factory-farmed products. Baur acknowledged that working at the federal or state level can be overwhelming, but at the local level people have a "bigger voice." Compassionate Communities is run by Nick Cooney, applying what he learned from writing his book *Change of Heart: What Psychology Can Teach Us About Spreading Social Change* (Cooney 2011) as well as survey findings from the Humane Research Council to encourage local activists to employ the most effective advocacy strategies in their interactions with their community and in Farm Sanctuary's design of some new short videos and leaflets.

Compassion Over Killing (COK) is also working on the supply side with consumers and the demand side with businesses. Their corporate outreach asks large retail/restaurant chains to put more (or at least some) vegan products on their menu, thereby reducing the amount of animal products the companies are buying as well as allowing all consumers wider access to vegan foods in the mainstream market. For example, COK is running positive campaigns to get a vegan donut at Dunkin Donuts and to get vegan meat (like a riblet or deli slices) on a standard sandwich at Subway, "the world's largest restaurant chain." And to encourage individual consumers to create demand for vegan products, COK is now running their 30-second factory farming commercials not just on cable television but also on *websites* where many younger people now go to watch popular TV shows online. Meier explained how airing it online increased hits to their website:

> The results were phenomenal. You know because they are already online watching this, people could directly click through to our website to get more information and request our veg starter guide. Within one week, visits to our website quadrupled and we were seeing a flood of orders coming in.

PETA also wants to make it easier for consumers to eat cruelty-free, even those people who say they never plan to stop eating meat. So Ingrid Newkirk pragmatically thought, "okay, then we'll create a meat that doesn't involve suffering and no animals will be used." So PETA has sponsored a contest offering one million dollars to the scientific entrepreneur who can develop commercially-viable in-vitro meat (flesh protein grown from cells in a lab, not on living animals). While Newkirk admits this isn't a "purist position," it works with America's appetite for animal meat without requiring anyone (human or nonhuman) to sacrifice. PETA is especially pushing for an in-vitro *chicken* meat, as "people eat a million of them an hour in the U.S. alone. We wanted to save these little birds that people don't often think about who suffer so terribly."

CORE CAUSES OF ANIMAL EXPLOITATION

When I asked what the real issue is that causes our society to exploit animals, Newkirk put it most succinctly in immediately replying: "human psychology, tradition, and habit." This summarizes many of the others' responses, as they tended to either focus on psychology (typically an insufficiency, such as a lack of honesty and self-reflection, connection, or awareness) or focus on behavior (bad habits based on a traditional use of animals). Hershaft noted it was hard to change the behavioral inertia based upon our tradition and history of eating animals. And Baur referred to these traditional behaviors as "bad habits," in particular our cultural practice of viewing and using other animals as commodities, failing to recognize life as sacred: "we've come to see certain things as normal, you know, bad has become normal." Baur also acknowledged the psychological aspect by saying that people tend to rationalize these cruel habits and make excuses because they are disconnected from their conscience and "better self." As emotional beings, when we feel badly about something, we often reveal ourselves more as a "rational*izing* animal than a rational animal."

This acknowledges that people *do* care about animals, a point Friedrich reiterated when he explained that the cause of human mistreatment of animals was cognitive dissonance – human denial about the reality of the animal abuse: "People love animals yet the interaction they have with animals the most is when they pay people to abuse them and slice their throats open." Therefore, he sees education as the answer. This is somewhat similar to Meier's response that it is just a lack of information that enables factory farming, and when people do find out about all the cruelty the industry hides from them, they are "outraged."

Friedrich later commented that most people believe animals are at our disposal, and I think this could be added to his hypothesis about psychological denial to imply that people may also be reluctant to question why they feel entitled to use other animals in the first place. This relates to Ball's response that speciesism, or as he put it, "unequal consideration of interests," *could* on one level be considered the most fundamental issue. But he thinks the most pressing issue to address is the "active exploitation of animals" – the institutionalized cruelty – regardless of whether people like or respect animals as individuals. Ball explained:

> You can just be simply human and repulsed by what goes on in factory farms and decide not to take part in it. And that I think really captures where we are at the moment in this country. We are so far from a culture of equal consideration of interests that to argue that we should be focusing on that very basic underlying principle will keep us from being as effective as possible at addressing the overt atrocities that are going on right now.

This sentiment is different from the thesis of my book that suggests we should directly confront the principles underlying our discrimination of others. Ball's views represent a more pragmatic emphasis on reducing the immense institutionalized cruelty. Perhaps this is just a matter of timing, as I note that the movement could do both simultaneously (perhaps privileging rights over welfare), or it could attempt to first reduce farming cruelty then work on ideological change decades later, once the culture has reached a tipping point in concern for farmed animal welfare and rates of vegetarianism.

Strategy

Beliefs about human nature and culture can play into the activists' broader advocacy goals and the strategy they use to achieve them. Ball said to reach Vegan Outreach's goal of reducing animal suffering, their strategy is to expose college students to factory farm suffering and provide them honest information about making more compassionate choices. Their *Why Vegan?* booklet cover is designed for people who are already interested in animal issues, but the *Even if You Like Meat* booklet is designed to appeal to the average college student when leafleting. It doesn't mention veganism/vegetarianism on the cover because psychological research suggests it's not effective to hit people with a "big ask" too soon (see Cooney 2011). He stated that the goal is to have people be receptive to taking and reading the booklet, and get people to start taking some steps toward a cruelty-free diet, hoping that once they identify with that compassionate choice they'll keep evolving. Ball explained: "We try to hit the sweet spot for reaching people where they are instead of telling them where we want them to be."

While Vegan Outreach takes a pragmatic approach while still prioritizing animal cruelty appeals, PETA's pragmatism extends to appealing to people based on "whatever reason moves them," as Newkirk knows that not everyone is altruistic. She explained:

> We can't simply say because *I* care about being fair to the other beings on the planet and relate to their suffering and want to stop it, I can't assume, because it's not true, that everyone will go vegan because of the reasons I went vegan or even that people who care about animals will go vegan.

She described PETA as more strategic than purist, focusing on "whatever will make people stop killing animals for food and stealing from them, their milk and eggs, at great physical costs to the animals." Sometimes this means using ethics-based appeals to get people to "relate to whose on your plate." Other times, it is just about introducing tasty vegan foods to people in fun ways, such as the "lettuce lady" giving out vegan hotdogs. She described:

> They slather them with chili and ketchup and relish and they have a great time and they don't really necessarily think about the animals who die in the meat trade, they simply think, 'Hey, Bill Clinton is a vegan, Russell Simmons' a vegan, Ellen Degeneres is a vegan, Venus Williams is a vegan. This is great! Someone is bringing me all this vegan stuff to try and boy it tastes good!'

This also attests to Newkirk's realization that the media has "regressed" into being more celebrity and soundbite-oriented than it used to be when PETA first started. It no longer allows for pockets of thoughtful philosophical discussion on animal rights but rather requires "more pictures, fewer words," and PETA's tactics have adapted to fit.

It is the media's lack of attention to factory farming that drives COK's two-part strategy to end animal suffering. It starts with exposing the truth about how animals are mistreated on factory farms and slaughterhouses. Meier said the next step is to channel the resulting consumer outrage by "encouraging people to stand up for these animals simply by not eating them" and giving people the "tools to empower them to make better choices." Part of this encouragement is to demonstrate how many people are starting to move away from meat, milk, and eggs. This capitalizes on the fact that psychologists tell us that people are more likely to follow along with something that is popular. Meier stated, "so I think it's really important to let people know that this is not a fringe issue anymore."

Similar to Meier, Baur also expressed a belief that people are "basically humane" and they just need to be made aware of why and how to "move toward plant-based eating." In support of Farm Sanctuary's goal of a vegan world, Baur described a three-pronged strategy based on the belief that "change happens incrementally." First Farm Sanctuary tries to change laws to prevent the worst suffering and raise awareness. Second, through shelter work, and "modeling a different kind of relationship with farm animals" Farm Sanctuary works to "change people's hearts and minds and re-evaluate our relationship with farm animals." And finally, they "promote compassionate vegan living," showing that "it's not a lifestyle of restriction but one of engagement." He shared: "Something I've been saying a lot lately is that we encourage people to make food choices that are aligned with their *own values* and aligned with their *own interests*." He acknowledged that he cannot tell people "how to think and how to behave," so he can only encourage people to voluntarily "want to be responsible for their food choices and want to be responsible for the consequences of their actions." Doing this is in their own interest, as otherwise the dissonance between one's values and one's actions "can cut into one's mental health, or emotional health, or even spiritual health and well-being."

When asked how protecting fish plays into their strategy, many activists acknowledged they should focus on them more. Hershaft compared the exclusion of fish to how the movement used to exclude chickens and focus instead on more loveable cows and pigs. This resulted in the public switching to eating more birds, implying this could happen with people switching to eating more fish. Meier agreed fish didn't get the attention they "deserve" from the movement. She explained that it can be difficult to articulate their individuality to people because "our government – the USDA – counts how many cows are killed for food, for example, but the individual fish are not counted. I think that's sort of reflective of our society's view of fish." Public sympathy for sea animals is less, partially because these animals seem less "familiar," as Baur put it. He suggested that to increase the familiarity of fish, activists should educate people about the similarities fish share with us and other animals (memories, pain, etc.). In this effort, Friedrich mentioned that Farm Sanctuary's new "Someone not Something" campaign includes a section on scientific studies of fish sentience. But Baur noted we should also highlight what makes fish special: "It's also good to celebrate differences and uniqueness of different animals that do things that humans can't do or other animals can't do. And that's another way to potentially gain some respect and appreciation for other species, including fish."

Ball was the only one who expressed a personal belief that many sea creatures (like oysters, clams, and shrimp) are not having as "morally relevant experiences" as birds and mammals are (in terms of ability to suffer), and therefore he preferred to focus his limited resources on reducing land animal suffering. He admitted that the farming of fish is an issue of concern as fish suffer for longer when farmed than when wild caught. Newkirk expressed the strongest concern for fish, mentioning PETA's campaigns to get people to realize it doesn't make sense for their compassion to stop at fish: "factory farming of fish is the same as factory farming of veal calves. You know, a fish feels the same as a kitten, so let's face facts." She thought the movement should focus on fish more, and also on the egg and dairy industry, as that is where there is the most animal suffering.

When I asked about how persuasive or important the environmental argument was to veg campaigns, most activists reluctantly admitted that the public did not seem as moved by it, even though the activists themselves felt it was very important and did include it in campaign materials. Hershaft said it wasn't as influential at getting people to make bigger, life-changing decisions such as what to eat. Ball agreed that, strategically, the environmental rationale was too impersonal, overwhelming, and less animal-centered to motivate people to stop eating animals, specifically. Similarly, Friedrich felt environmental rationales were not as "viscerally" compelling as animal cruelty or personal health rationales at getting consumers to go veg. To make a dent in our environmental problems requires systemic not just individual change, and he expressed a concern that the United States lacked the political will to make the significant environmental policy changes necessary, especially as many conservative politicians won't admit that humans are causing global warming. Yet he noted, "we *can* get conservatives on the humane argument." Baur felt that the environmental argument for vegetarianism was something that the *environmental* movement should take on. And Meier agreed that, while it was part of her "toolkit," the argument would seem more credible coming from environmentalists, rather than animal activists, as that is their area of expertise.

When I asked activists about the challenges and importance in American culture of animal rights activist comparing humans to other animals, they all admitted it was a challenge. It could be addressed in a variety of ways:

1) comparing our cognition and capabilities,
2) suggesting our interests should be considered equally /fairly, and
3) comparing types of exploitation experienced.[2]

Friedrich chose to emphasize the first way, by citing scientific studies that prove some other animals experience similar abilities to feel pain and emotions as we do. Baur liked to highlight how we are more similar than different, saying it was useful to focus on "the commonality that all of us want to live, we don't want to suffer, and we want to enjoy life." But he said it was also "legitimate" to question how we humans are different from other animals, and "just because some are different, though, does not mean that some are better or worse." Friedrich felt it was "important" to make these scientific arguments especially to "influentials" in forums where "we have the capacity to have a real and rational discussion with people."

But when it comes to appealing to the average American, activists tended to agree that the dog and cat comparison to farmed animals is more strategic or persuasive than the human comparison to farmed animals. Strategically, Meier, Ball, and Hershaft all acknowledged that people often get "defensive" when they feel they are being compared to "animals;" therefore, that can distract from them hearing your point about farmed animals or veganism and gives them an "excuse to ignore the rest of your message" (as Ball put it). These activists admitted on some level it is important to make the argument that humans are animals, as not everyone believes that, but it isn't as "easy" or "powerful" (to use Meier's terms) as comparing nonhumans to other nonhumans.

---

[2] The second point about fairness was not directly addressed by the activists when I asked this question, but I included it because I think it is relevant and also because it links points one and three, points which activists *did* directly mention when interpreting my question.

Newkirk's response to my question was the most unique, as she interpreted it in terms of the third type of comparison between humans and other animals – that of comparing systems of exploitation such as the Holocaust, slavery, and factory farming. She immediately responded "Well, of course we've certainly take it in the neck for daring to compare one group of victims to another group of victims, because victimization is very precious to many people and they can't get beyond the group they relate to." Newkirk saw the comparison of human-animal exploitation systems as logically consistent, explaining: "As long as it's put in the right context, then surely one should still be upset about others whose lives consist of being in chains and being dominated and being enslaved to human beings. That should be fairly easy." Yet while PETA does not mean to be disrespectful to any victims of exploitation, there has been backlash against PETA campaigns, particularly related to race (see Kim 2011; Freeman 2007). In PETA's defense, Newkirk stated:

> There will always be peculiar particular things that happen to any individual or groups of individuals. But the experience that is important is that it was exploitation, discrimination, de-valuing of life, it was a horrible failure to relate. And yet the challenge in posing this is that the inability to relate is still there [to nonhuman animals].

I acknowledged that perhaps people were still not ready to be compared to other animals, but asked if she felt the comparison was necessary to make because this separation between us is part of the problem. Newkirk responded: "It's the *whole* problem. If you can't relate, then you're going to discriminate."[3] While the other activists tended to want to avoid public scrutiny, feeling it wasn't as effective for the animals, Newkirk felt it was necessary to face this ridicule, as feminist and abolitionist Sojourner Truth did. Newkirk declared:

> It can never be too early to start an idea that will reduce suffering or result in less discrimination because if you wait until everybody's ready for it, then there's no point in saying it… You have to just carve away. Stand up and say it.

Regarding my earlier note about timing and movement phases for different types of messages, Newkirk's comments suggest that we

[3] We also discussed how this rhyming phrase would make a great protest message or t-shirt slogan.

shouldn't wait to ask for animal rights, and it can be done simultaneously with campaigns aimed at reducing cruelty. When I had asked how PETA would like people to view humans in relation to other animals, Newkirk replied:

> My dream is that one day we'll shed all these prejudices that we have and start not to see ourselves as demigods but see ourselves as simply one animal among many… We would start saying "I understand I must be considerate to you not because you are a human being but because you are a *living* being."

VALUES

When asked what key cultural values are most important to appeal to in campaigns, the activist leaders' most popular answer was *compassion* and *kindness*. Parallel to this is the value of *integrity*. While the activists didn't mention "integrity" specifically, I highlight it as implied because integrity is based on displaying *moral consistency*. And activists frequently emphasized the desire for people to consistently act upon their compassionate values – not only *saying* you are compassionate toward animals, but actually *being* compassionate to all sentient animals, including farmed animals. For example, Meier stated: "the number one key value is encouraging people to *express their compassion* for all animals by not eating them." And Friedrich implied integrity to me in this way:

> The vast majority of Americans – 97% percent – think that abusing animals is wrong and should be illegal. And yet about 97% of Americans also pay people to abuse animals for the inconsequential good of a palate preference. So we don't really have to change people. We don't have to convince people to change what they already believe, by and large. For the vast majority of people, we just have to help them align what they already believe with their actions.

Newkirk described this approach as trying to "appeal to reason" by asking for consistency in applying their principles of compassion and *nonviolence* to all, saying:

> We try to spread governing life principles that people have in a reasonable, logical way, so that they can understand that there are others that want to come under their cloak of understanding, compassion, kindness, consideration, respect, and so on.

At the core, here, is the activists' belief that people value self-identifying as a "good person" (although, Ball was the only interviewee to use that phrase). If campaigns point out ways that people's actions are not in line with their values, it is meant to challenge people's sense of self. Ball doesn't believe that people necessarily identify with being *compassionate* toward animals, or loving animals, or thinking animals are equally valuable, but there is a stronger identification with wanting to *avoid* feeling like a bad or mean person who harms animals. Basically, while they may not be animal lovers, they *aren't animal abusers*. He explained: "We can *respect ourselves* and the values that we hold regardless of our feelings about other animals, and not want to be a part of a system that we recognize as inherently cruel." This is why Vegan Outreach booklets highlight cruelty as the main problem.

Ball's focus on "people's revulsion to cruelty" aligns with what Meier and Hershaft implied about human motivations when I had asked a related question about whether we should appeal to the value of *freedom* for animals in our campaigns. They implied that animal rights is not so much about caring for other animals (and their freedom or rights) per se, but mainly about caring for *ourselves* and maintaining a respectable cultural identity. Meier explained that rather than focusing on "what to *give* an animal," COK focuses on "how we are *harming* animals, because somebody can relate to that" since it is "simply out of step with our society's values to treat an animal like this."[4] And Hershaft said that my focus on freedom was really referring to animal rights, something that ironically "really has not been addressed very much within our movement." He explained his view that: "The way to address animal rights with people is to explain that it's not about the animals. It's about *us*. It's about how we relate to other living beings." And we needed to decide how to relate to them "in a manner that allows us to be consistent with our own moral values." In this way, I believe many of the activists' comments collectively show we must ultimately appeal to a sense of moral integrity at an individual and societal level.

The notion of integrity is related to demonstrating social responsibility, although that wasn't a phrase I heard the activists mention

---

[4] Another way of phrasing Ball's and Meier's comments is that animal rights is more popular when portrayed as a negative right (don't do something harmful) rather than a positive right (you should do something good).

much. However Baur listed responsibility as a key value by saying: "We want people to act *responsibly* and to not cause harm to others, or to the planet, or ultimately to themselves." Regarding the latter point, he noted that because kindness toward animals is "also good for people," Farm Sanctuary also sees itself as promoting human "health and wellness more broadly," both in a physical and mental sense. I see moral integrity as closely aligned with mental health and well-being.

Defining the Problem with or as Animal Agriculture

While factory farming obviously comprises the majority of farming practices, I asked activist leaders how important it was to problematize *all* animal farming (even small or free-range farms) so that the emphasis was specifically on critiquing the *use* of someone as a resource rather than just treatment. Friedrich said Farm Sanctuary generally critiques both unless they are just focusing on legislation. The new campaign championing the fact that farm animals are "Someone Not Something" ultimately "transcends the abuse and simply says no matter how well a dog or a cat is treated, you're not going to eat her… yet no one can come up with a rational explanation for how there's any difference between eating a cat and a chicken, a dog or a pig." Baur said, "as soon as an animal is seen as a resource there is going to be a tendency for there to be problems." This can happen on smaller farms too. Farm Sanctuary defines *factory farming* more broadly than the public would, describing it as:

> A commodification of sentient life and the perception of animals and the natural world as resources to be exploited. And so if you have animals and even if they live well and they are killed to be eaten, that is essentially the commodification of sentient life.

Baur noted that "killing is inherently violent." Meier also emphasized the killing inherent to all types of farming: "free range animals or grass fed beef. These animals might lead slightly better lives than their brethren on factory farms, but at the end of the day they are still slaughtered for us to eat." She stated:

> The reality is that when we choose to eat animals, they are killed for us to eat. And I think that is an important aspect to focus on. No matter how that animal was raised, we are making the conscious decision to say "yes kill that animal so I can eat the flesh."

In discussing smaller farming, the activists used astute terminology that acknowledged some benefits while still framing it in a negative way overall as less bad, less cruel, or less atrocious, more often than they used industry terms like "humane farming." For example, Newkirk noted that offering more space was a good step, but it still denies these animals too much, so "the key is that it is called *less cruel*. It's not humane." Similarly, Meier affirmed, "it causes less suffering to the animal, but it is *not cruelty-free*." And Baur felt that: "if somebody decides they are going to eat meat, it's certainly better to do so from farms where the animals are treated *less badly*," but people need to visit these farms, as often "the conditions are not nearly as good as people assume they are… it's better, but it's still *not good*." Powerfully, Hershaft described these farms as using "*less atrocious* methods than factory farming."

Hershaft was discussing both the philosophical and the strategic way to respond to the trend toward consumers embracing "less atrocious" farming. Strategically, he thought it is ultimately unsustainable because it cannot produce "anywhere near the amount of animal products" that the public demands, and the price would "skyrocket," which might ultimately "destroy the meat product system" in the long run. From that perspective, the animal rights movement could see organic or free-range farming as a beneficial trend. But from a philosophical/ethical perspective, he saw it as a "dangerous trend because it takes a lot of low hanging fruit – all the people who are most likely to become vegans for ethical reasons – and it transforms them into 'ethical carnists,'" taking some of the vegan movement's "prospects."

Ball said that for general outreach, Vegan Outreach focuses most of its attention on factory farming and is less concerned about the trend toward so-called "humane meats," as it probably isn't as popular with the general public as vegan activists think it is:

> It's too expensive and it's too rare. It's too hard to do, and it's something that people won't be able to do unless they are rich and they have lots of time on their hands. And that's just not our main focus. Our main concern – Vegan Outreach's main concern – isn't that people eat humane meat, it's that people stop eating big animals and then eat more small animals. They are causing a lot more suffering when they do that.

In utilitarian fashion, Vegan Outreach wants to save the greatest number of lives, thereby reducing the most suffering. In America, this is done by saving the smallest, most mass-produced (and mistreated) animal used for meat – birds. In their meat reduction messages, Vegan Outreach has started encouraging people to give up birds, in particular. Ball is trying to prevent the trend of people responding to animal cruelty, environmental, or health messages by giving up red meat and switching to chicken. For that reason, in Vegan Outreach's booklets, there are many images and descriptions of birds and much less about cows.

MEAT REDUCTION *VERSUS* VEGANISM OR MEAT REDUCTION *TO* VEGANISM

I posed the question that, even though environmental and health groups might see no problem with asking people just to *reduce* their meat intake, how does an animal rights organization decide whether to ask people not to eat *any* animals or just to eat *fewer* animals. All of the activists rejected my dichotomizing of the issue, to some degree, and saw meat reduction and veganism not as separate categories but as paired along a progressive spectrum, with veganism at one end. For example, Hershaft responded: "We don't see a conflict. We basically feel meat reduction is an incremental step toward veganism. And we always remind people that veganism is the ultimate goal." Meier also described reduction as a "step" along a "path with the ultimate goal of not eating any animals." While COK asks people to "express compassion for all animals simply by not eating them," she acknowledged that it's "not an immediate issue in terms of all at once, like if you don't go vegetarian or vegan today, then just forget it." She showed understanding for people wanting to take it slowly, saying "that's okay, but just try to incorporate more vegan meals into your diet and just try to do that a little more every day as you go down this path."

Newkirk also used the footsteps metaphor, saying: "there has to be a step-by-step process," as you "can't expect to veganize a school with the snap of your fingers." However, she clarified that "without exception, we say ideally you could just go vegan." She also framed this process using a swimming metaphor, stating, "we certainly do try to get that toe in the water, then the foot. And before they know it, they are swimming." She comically explained this progressive process:

> We try to introduce meatless, dairy-free, eggless meals to people and say "just try some of them" because if they reduce their meat consumption by one meal, that's

> a good thing – by one day, that's an even better thing. But rather than Meat-Free Mondays, Meat-Free *Sundays through Mondays* is even better. [she laughs] And the ultimate goal of course is dairy-meat-and-egg-less Sundays through Mondays.

Farm Sanctuary takes a similar incremental approach, which Baur described as "pragmatic" but also "aspirational." He explained:

> Our goal is to be effective, and if asking them to eat fewer animal foods is more effective, that's, I think, the correct approach to take. But we also always put out the vegan aspiration and do not shy away from the reality and violence of slaughter.

He declared that it's important to be "clear" about the ultimate goal being a "vegan world," while also recognizing that it is process that will take "a long time, and whether or not it's even possible is kind of an open question." The vegan world aspiration is part of what makes the dietary message of animal rights organizations different than that of other causes. For example, Friedrich noted that:

> Our end goal is veganism from an animal's perspective; certainly the environment and global poverty movements should have the same goal, just for environmentalism and food crisis reasons. They don't, for reasons that I'm sure are more tactical than philosophical.

Friedrich surmised it might be dissonance over the fact that many of the environmental movement and anti-poverty movement leaders may be meat-eaters and also that "they look at the landscape and look at their major donors list and say 'we can't ask for more than this.'" But he acknowledged it is strange for them to acquiesce on the food issue, given the gravity of the situation, as it would be odd to hear environmentalists ask people to avoid driving gas-guzzling Hummers just on Mondays, the way they settle for Meatless Mondays. But this doesn't mean Friedrich is against Meatless Mondays, as from a practical standpoint, he pointed out that if every American participated in Meatless Mondays it would be the equivalent of $1/7^{th}$ of the population essentially going vegetarian – about 45 million people.

Vegan Outreach's *Even if You Like Meat* booklet is aimed at getting people to take the "first step" of reducing their contribution to animal cruelty by eating less meat. In this way, Ball reiterated what many of the activists said in terms of believing that people's diets will evolve toward veganism over time if they would just take that first step, especially if it is motivated by an animal cruelty frame. He explained:

> Once they've taken a step of reducing meat, then they are saying this is a value important to me, of not contributing to this animal cruelty, which is why I'm doing this. And so the connection has been made, since we've opened the door, this will continue to evolve in their mind.

This represents a belief in a "foot in the door" spillover approach to behavioral change (Cooney 2011). And Ball noted that, culturally, vegetarianism is more prominent now than it was decades ago, so that also reinforces the chances that people would evolve to further reduce their meat intake rather than sliding back into their old habits. But he also looked at my question from a different angle and explained that it is "hypocritical" for activists to insist on veganism as the only stage of the dietary spectrum that is moral:

> As an animal rights group, it's easy to argue that animals have the right to live, so you have to tell everyone that animals have the right to live and *any* taking of their life is wrong. But in reality, all of us are taking the lives of animals in some way or another through our choices: we drive we hit animals, we buy food that was harvested and transported and the process of that kills animals, and all those animals had a right to live as well. And if we recognize that, it's not only not psychologically effective in terms of advocacy, but it's also hypocritical for us to make claims that people have to act a certain way [veganism] or else they are morally wrong. When we are doing the same, you know. Our motivation might be different, but the actual results are different only by degree – a significant degree, but they are not different in kind.

While it is good to acknowledge that there is not a 100% cruelty-free diet/lifestyle, I think that all the activists' responses, including Ball's, indicated that they feel it is *better* (morally) to do *more* to intentionally reduce suffering and save lives than it is to do less. I conclude this because all the activists hoped that people would evolve toward a vegan end goal. So while veganism may not be a perfect diet, it is framed as being the best or ideal diet.

FUTURE VISION FOR FARMED ANIMAL PROTECTION

NAME	What CONCERNS You about the Future?	What Makes You HOPEFUL about the Future?
Ingrid Newkirk (PETA)	Segmentation / fracturing of people's media consumption making it harder to reach everyone	Multiple factors pushing toward a shift to a vegan diet (public health/obesity, environmentalism, and animal protection)
Alex Hershaft (FARM)	Less excitement now in grassroots animal activism	Globally, society will eventually have to move toward plant-based for environmental and hunger reasons. Reaching a critical mass in public awareness of the healthfulness of a plant-based diet. Plus animal activist tactics are now more effective.
Matt Ball (Vegan Outreach)	Elite/purist veganism being counterproductive. Plus people switching from eating larger animals to eating (more) smaller animals – birds.	Animal activists are more effective now and public concern is growing. Trend is toward developing mainstream veggie versions of meat, egg, and dairy products (alternatives). Economic issues will also push a shift to a plant-based diet.
Bruce Friedrich (Farm Sanctuary)	Counterproductive elite/purist vegan advocacy is a small concern	There is momentum behind the movement. National animal organizations are effectively advocating for farming reform and veganism.

Gene Baur (Farm Sanctuary)	Rising trend toward meat-eating and factory farming in developing countries	Public has greater awareness and access to info to make compassionate choices. Growing food movement reveals a convergence around animal protection, health, and environmental issues. Animal farming is economically inefficient.
Erica Meier (Compassion Over Killing)	Rising trend toward factory farming in other countries	Meat consumption is declining in the U.S., people are talking more about veganism, and factory farming practices are being challenged and worst ones are phasing out.

My last question for the activists was to have them share their projections for the future of animal advocacy, taking into account global trends: what made them concerned, but also what made them hopeful. Ball and Friedrich were optimistic about how animal rights activism had progressed to be increasingly effective and pragmatic. Yet they both shared a concern that some vegan activism is "counter-productive" and divisive. Friedrich explained it as vegan activist "puritans" who would not support fellow animal organizations if they work on reforms and not just abolition. And Ball referred to it as "elite veganism," also adding in a health-purist component. He described the mentality as: "everyone has to eat exactly what I eat and has to think exactly what I think and you have to have an all whole foods, locally-grown organic diet." He said that was "not going to cut it" anywhere besides liberal cities like San Francisco, as most Americans don't want to eat that differently. This is why he was so optimistic about the trends towards mainstream marketing of vegan versions of meat, egg, and dairy products – foods that people like, are familiar with, and want to keep eating.

When talking about activism, Hershaft felt the problem wasn't activists being too ideological or purist, but rather a "loss in grassroots activism, in excitement, in leadership within our movement," as there has been an "institutionalization of activism" into larger animal protection organizations. But, despite "less grassroots activity, less drama in our public posture," he was impressed that animal activism is now "better financed, better thought out" and is "reaching more people and being more effective."

Yet Newkirk expressed a concern about the challenging micro-media environment that makes it "even harder to reach people unless they already care about animals." This is because of how many different and specialized media options people have where they can sequester themselves in their own bubble. This explains some of PETA's controversial sex and nudity-oriented campaigns that attract wider attention but offend some who see the campaigns as sexist or indecent (see Deckha 2008; Pace 2005). Newkirk defended this tactic by saying:

> If you look at the stats, like it or lump it, more people are watching things about sex than anything else on the internet, so our obligation is to make sure that they also hear something about vegetarian/vegan diets while they are watching sex.

But Baur noted a flipside of the modern media environment as making it easier to "reach more people more quickly, more inexpensively, with the internet and social networks." Based on all this communication technology, he felt "there's more awareness now than ever before." And Meier expressed pleasure at how, especially in recent years, the public is more frequently hearing about veganism from mainstream sources such as the *New York Times* "encouraging people to think about their food," which she saw as "progress."

While Baur and Meier expressed some similar optimism, they also had some shared concerns – concerns that factory farming practices and a Western meat-based diet were being exported to other countries. Baur stated: "I'd love for the best qualities of America and of developed countries to be adopted, but unfortunately it seems that some of the worst qualities are being adopted and are perceived somehow as good." He blamed this on issues of power, as meat has been perceived globally as a symbol of wealth, progress, and power – "power over somebody else." In this way, meat "feeds the ego." He hoped that these other countries will consider their own cultural values and the negative consequences of factory farming and learn from our "mistakes."

Hershaft is concerned that it will get worse before it gets better, and the world will see an increase in nonhuman and human suffering (such as with malnutrition and hunger) before we realize that we "defacto *have to* turn to a plant-based diet." He predicted this transition is inevitable because the present methods of raising animals are "obviously unsustainable." This is comparable to Ball's and Baur's comments that the economic inefficiency of animal farming will hasten its eventual demise. Similarly, Newkirk predicted: "I think the future of the world, it *has to* go vegetarian and vegan." She felt that as people are recognizing all the problems with factory farm pollution, human disease and obesity, in addition to the "humane concerns – the horror of it all" – these factors all mitigate in favor a change. Baur agreed that with the growing food movement and concerns about factory farming, there is a "convergence of issues" now with environmental, animal welfare, and health issues. Hershaft is particularly optimistic that society has reached a "critical mass" in recognizing the health risks of eating animals, where at least 20% of the population acknowledges the health benefits of vegetarian meals and change should proceed rapidly.

Overall, the activists' concerns were not as deep as their optimism, as many saw the movement as headed in the right direction with a lot of momentum. As Friedrich put it, "the winds are at our backs." And Baur predicted: "Just as cruelty can become normalized and spread, so can compassion, so can sustainability, so can healthful living. And I think that is starting to happen." Meier was equally optimistic:

> From factory farming issues slowly being challenged and some of the cruelest practices being phased out to an increasing number of people choosing to leave meat, milk, and eggs out of their diet, we are seeing this increase and it's all happening, just even in the last five or six years, we are seeing such huge shift.

Since Meier believes that "progress begets more progress," she told me: "We are going to keep pushing forward and next time, maybe in four years,[5] when we talk again about these issues, we'll have even more good news to highlight and tout."

I'm not sure I would call myself an optimist, but the mere fact that I work so hard to convince people to stop eating fellow animals reveals an innate optimism that this is a legitimate goal and that my (our) advocacy makes a difference. In the next and final chapter of

---

[5] She's alluding to the fact that I first interviewed her and all these activists four years ago, in 2008.

this book I share my specific recommendations for how animal activists should frame the vegan issue. My recommendations can be considered in concert with the advocacy recommendations provided here in this chapter by the six prominent vegan activists I interviewed. Even though our opinions on strategy may differ at times, we all share the similar goal of progressing to a vegan world with less suffering and exploitation. Toward this end, Friedrich puts the goal in positive perspective:

> I think we'll get there. I take great solace in thinking about the fact that if we had been having this conversation just 150 years ago we might be talking about whether slavery was natural for human interactions, and whether women have brains that were developed enough that they could rationally consider political problems and have a say in governance. And science was able to allow us to get past those significant forms of bigotry. I think we'll do the same thing on animals. It's interesting, 150 years is a finger snap, right, relative to even civilized times. Socrates said, "the unexamined life is not worth living," 2,500 years ago. For 2,350 years we held these views we now recognize to be diametrically the opposite of what we understand to be true morally and ethically today. So it's a process, but I think it's quite possible that we'll move through it with regards to our biases against animals fairly quickly.

Chapter 9

Recommendations for Ideological Authenticity in Framing Animal Rights

In this book I discussed communication strategy with the leaders of premier United States animal rights organizations working on national food and farmed animal issues, and I defined and categorized the main frames they use in their persuasive messages. To structure my findings, I used Snow & Benford's (1988) collective action frame categories as a guide, starting with diagnosis and prognosis. My study revealed that organizations' diagnosed the problems as: 1) suffering of animals due to cruelty; 2) commodification of animals into economic objects; 3) harmfulness of animal agribusiness and animal products to humans and the environment; and 4) the needless killing and death of animals for food products. As part of these problem frames, animal advocates largely blamed animal agribusiness, and to a lesser extent, the fishing industry, for causing cruelty and destruction and hiding it from public view. Animal rights organizations sometimes made American consumers a secondary responsible party once these consumers were informed of problems associated with an animal-based diet.

The most popular prognosis was for organizations to suggest that consumers eat fewer or no animal products. Farm Sanctuary also promoted animal welfare reforms by government and PETA also promoted some welfare reforms by industry and meat retailers. This reflects some variance in terms of Benford & Snow's (2000) notions of flexibility and inclusivity within frames, as some animal organizations were more flexible in their recommended dietary changes for consumers and in whether or not they included industry and government as part of the solution.

In studying the prognosis and diagnosis framing components, I also identified the major *values* to which animal organizations appealed, inspired by Snow & Benford's (1988) motivation component of collective action frames and Gamson's (1992) similar identity component. These motivational values were: *compassion and caring for nonhuman animal suffering and an aversion to cruelty; respect for the sentience and individuality of other animals; moral integrity and consistency; desire to improve the world and make a difference; choice; pleasurable and convenient food; belonging; life; naturalness; honesty; concern for fellow human beings; American populism and accountability of big business and government to the people; freedom;* and *American pride.*

In framing problems and motivating change, activists often used animal *welfare* ideology to achieve animal *rights* solutions. If frames can be perceived as a recruiting tool for ideologies (Oliver and Johnston 2005), then these animal rights organizations are primarily recruiting based on an expanded notion of animal welfare ideology more so than rights. Therefore, in this final chapter I'll recommend framing strategy that would arguably create more alignment between theory and practice, specifically, better aligning animal rights ideology with the organization's communication strategy (Baker and Martinson 2001; Francione 1996; L. Hall 2006a; Lakoff 2004; LaVeck 2006b). Throughout, I'll share some specific rhetorical examples of current or potential messages that fit my recommendations.

To summarize my perspective on the frame alignment process, the ideal frames for animal rights organizations to use would be ones that are truthful as well as congruent with an animal rights ideology (deontological), first and foremost, as well as being effective at meeting animal rights goals (utilitarian). I advocate for what I refer to as *ideological authenticity*, meaning what is true or authentic to a social movement's ideology should be expressed as such, in most cases, for integrity and honesty in communication. As animal activists are part of a challenging movement that seeks a fundamental transformation in worldviews and behaviors, I advocate for some frames to fit a frame *transformation* alignment process (Snow et al 1986), in support of Foucault's (2000) notion of critical transformation and Lakoff's (2004) idea of reframing issues for social change. To do this, farmed animal rights organizations must ask the public for the kind of major change in speciesist worldview that is necessary to promote all animal rights issues in the long-term while still finding a way to resonate with the public. This chapter explains my ideas for how animal rights advocates could construct less speciesist frames that resonate on some level with a largely speciesist public.

First, I discuss my recommendations for framing animal foods as a problem based on 1) injustice, 2) cruelty and suffering, and 3) environmental destruction. Then, I suggest engaging the audience as both consumers and citizens to explain their culpability and their capability toward individual and collective solutions. These solutions include: 1) appreciating the mutual subject status of all animals (including humans); 2) eating a plant-based diet; and 3) working collectively to create a less speciesist society. Infused through all of this should be appeals to values such as: *fairness, respect, life, freedom, integrity, honesty, naturalness, vitality, responsibility, moderation, community, diversity, caring, compassion, peace, sharing, humility, accountability, making a difference, self-esteem, health,* and *personal growth/ development.*

CHART OUTLINING MY FRAMING RECOMMENDATIONS FOR VEGAN CAMPAIGNS

FRAMING PROBLEMS		**RESPECTIVE VALUES TO PROMOTE**	
1. Injustice toward nonhuman animals through exploitation and unnecessary killing caused by animal agribusiness and consumption of animal products		Justice Life Respect Naturalness	Freedom Animal Dignity Integrity
2. Cruelty and suffering inherent in farming anyone *[cruelty is a subset of the injustice frame]*		Caring Life Fairness	Empathy Compassion Peace
3. Destruction of the environment and harm to wildlife caused by raising animals for human food		Fairness Moderation Community Diversity Vitality Sustenance/Sustainability	Responsibility Sharing Interdependence Naturalness Life

FRAMING CULPABLE PARTIES		**RESPECTIVE VALUES TO PROMOTE**	
1. Meat-eating individuals and a carnistic culture		Honesty Integrity Caring Self-Respect Self-Improvement	Accountability Harmony Peace (of mind)
2. Animal agribusiness and fishing industries			
FRAMING SOLUTIONS		**RESPECTIVE VALUES TO PROMOTE**	
1. Respecting the sentience and interests of fellow animals, and embracing our own animality		Respect Connection Fairness Humility	Kinship Diversity Empathy Self-Esteem
2. Eating a plant-based, largely organic diet and supporting responsible food choices in the community		Health & Wellbeing Life Sharing Harmony Belonging Choice Making a difference/ Usefulness/ Importance	 Vitality Community Responsibility Satisfaction Naturalness
3. Working collectively to protect other animals from exploitation and to create a less speciesist society – a just humanimality		Respect Fairness Democracy Growth Faith/Hope	Responsibility Community Altruism Integrity Empowerment

RECOMMENDED PROBLEM FRAMES

Injustice

I believe that, when advocating for animals used for human food, animal rights activists should primarily problematize *injustice* more so than problematizing suffering on factory farms. An injustice frame would be transformational in nature, asking people to question carnism and reconceptualize the accepted practices of animal agriculture, fishing, and meat-eating as unacceptable practices on the basis that they are, in most cases, unjust and exploitative. This frame could be complemented by promoting values of *respect* for the *lives* of sentient animal subjects, and respect for the *dignity* and *freedom* these animals deserve as capable adults.

The injustice frame should incorporate ethical aspects drawn both from nature *and* human culture. To include the latter, we could highlight human culture's appreciation for justice and fairness by protecting the lives and rights of fellow subjects not to be unjustly killed or exploited by our own species. To incorporate principles from nature, we could show appreciation for the right of all animals to live free from control and domestication, while acknowledging that healthy ecological life cycles contain some necessary violence.[1] The injustice frame should state that animal agriculture is unfair and unnatural because it includes breeding someone in captivity, growing him/her to suit one's needs, and exploiting his/her body and young for one's own benefit. The exploited individual does not have the natural opportunity (nor perhaps the life skills) to leave the situation and survive on his or her own, nor the freedom to own his/her body and control what is done to it. As an example of this frame, I like how vegan lecturer Gary Yourofsky (2010) put it: "How would you feel if the day you were born someone else had already planned the day of your execution?" And a basic way I like to phrase this sentiment is: "it's unfair to farm someone."

The practice of *hunting* animals for food is not as easy to fit into an injustice frame as the practice of *farming* animals is, especially a frame reliant on the corresponding values of *naturalness* and *freedom*. Hunting becomes more natural and more justified if animal products are required for survival, as they are for human animals in some parts of the world, and as they are for wild omnivorous and carnivorous nonhuman animals. Hunting does not involve the captivity and lifelong ownership that agriculture does, so it is less associated with exploitation and enslavement. Therefore, I contend that fishing by humans, as a form of hunting, primarily becomes unjust if the capture and death of sea animals is not necessary for basic human survival. To

---

[1] In my moral vision for a just humanimality, we don't interfere with natural predation among nonhuman animal species, although we might become predators ourselves if our lives depended on it. We should not domesticate anyone; nonhuman and human interactions should be mutual and on equal terms. But if there are wounded or orphaned nonhuman animals whose lives and existence could be improved by our care, then we can provide it.

the extent that humans can survive on plant-based foods and any necessary supplements, they should do this to avoid intentional killing and unnecessary violence per cultural values (at least as they are applied toward human subjects).

Vegan campaigns rarely challenge farming on the basis of it being unnatural, except they sometimes say that it's unnatural to drink the milk of another species. To question the naturalness and biological necessity of farming itself, I can envision a t-shirt/poster with a picture of a carnivorous species (such as a lion), with the saying "She *has to* eat animals to survive. We don't. Go Vegan." Or it could read "You don't see her farming antelope. All animals deserve a chance. Just farm plants." Another way to imply naturalness in a vegan message is to scientifically classify oneself as an "Herbivore," as do some of the messages from the Herbivore Clothing Company.

Because I don't demonize predation or rule out the possibility of hunting being ethical in some circumstances, my viewpoint incorporates flexibility around notions of "necessity" in terms of when violence is needed for survival. But it is better to have the debate center upon determining what is fair and necessary in the basic *killing and consumption* of other animals rather than continuing to center the debate on whether certain animal *husbandry practices* are inhumane. This follows animal activist suggestions that the animal rights movement should control the discourse around the problem of exploitation rather than husbandry practices (Dunayer 2006; Francione 1996; L. Hall 2006a; LaVeck 2006a). Therefore, I'd like to see activists increase their use of the frame that problematizes the killing and taking of life for human food by labeling the violence as *unnecessary, needless, pointless,* or even *gratuitous* (refer back to chapter 4).

To help demonstrate the monstrous scale of the killing in an engaging way, FARM's homepage now has a ticker that tells viewers how many birds, pigs, cows, and sheep the industry has killed worldwide just since the viewer opened the webpage. The numbers are whizzing by faster than you can keep up with, with approximately 1,000 chickens killed per second. And to emphasize how each animal's death is premature, Farm Sanctuary's "Humane Myth" brochure tells readers the natural lifespan of each species and at what surprisingly young age each species is killed by industry. It becomes obvious that we are often killing juveniles. And to remind people that the wild ancestors of farmed animal species do get to live freely and have the chance to live out their natural lifespans, activist materials should show fully-grown wild turkeys, boars, bison, big-horn sheep, ducks, pheasants, jack rabbits, fish, lobsters, squid, and other similar species in their native habitats, surviving as dignified adults and parents within their social groups/families. Showing these "wild" species helps emphasize naturalness and freedom as part of the moral vision of animal rights.

I believe the activists' appeal to what I called "American populist" values, which promote siding with the "little guy" against elites and corporations, fits better with the activists' problem frame of corporate farming cruelty than it does with my proposed injustice problem frame.[2] While this may seem counterintuitive because American populism *is* a pro-justice frame promoting egalitarianism, I think its reliance on anti-corporate values *implies* that smaller or "family" animal farms are unproblematic because they benefit the middle-class farmer and treat human workers and other animals fairer. The injustice frame, as I am recommending it, is not specifically anti-corporate as much as it is *anti-exploitation, anti-enslavement,* and *anti-violence*, whether the perpetrator is a corporation, a family farm, or a single person.

Some examples of messages that fit the injustice problem frame include:

- The lesser-used classic slogan "Meat is Murder" (PETA sometimes uses this).
- Be a cereal killer not a serial killer. Eat animal-free foods.
- A fundraising mailer from Farm Sanctuary that features a pig behind bars on the envelope with the phrase "How many babies have been stolen from her?" This fits with any messages that emphasize that milk, eggs, and babies belong to their mother, not a man or a corporation. Other ways to put this would be "AVOID STEALING what belongs to animals. Eat Vegan Instead," or "Vegan: nothing taken from an animal."
- FARM's new teen website FarmUnderground.org promotes abolitionism and has a justice-oriented leaflet called "Break the chains of slavery," featuring a chain breaking in front of a chick. The text on the back uses terms like *oppression, rights,*

[2] The populism frame may be useful during the limited anthropocentric altruism appeals that highlight the harm caused to humans by modern farming and fishing.

and *exploitation* to explain how we treat farmed animals like commodities, relegating them to property status. It is titled "Challenge society" and the call-to-action reads "Refuse to subsidize slavery. Live Vegan!"

- PETA's Chew on This DVD lists some injustice-related moral rationales for veganism, including: "because no animal deserves to die for your taste buds," "because commerce is no excuse for murder," "because might does not make right," "because they're defenseless," and "because it isn't fair." And PETA's belief button directly expresses animal rights: "Animals are not ours to eat, wear, or experiment on."

If "injustice" seems too indicative of criminal justice and legal systems, the term "unfairness" gets to the heart of what the problem with farming really is.

Cruelty and Suffering

Animal activists framed problems around cruelty more so than any other issue; in chapter four I explained the trouble associated with primarily emphasizing a cruelty frame, so I maintain that it should be used selectively. But there are certainly aspects of it that are in alignment with animal rights, as it shows a concern for nonhuman animals as sentient individuals who are equally interested in avoiding pain and suffering as are humans. Therefore, it is best used to complement a solution that asks humans to see other animals as fellow persons and to value their sentience and individuality so that humans avoid treating them like objects or causing them to suffer.

A key challenge with the cruelty and suffering frame is that it usually constrains the discourse to a debate over animal welfare within farming rather than debating the necessity and fairness of farming itself (Dunayer 2006; Francione 1996; L. Hall 2006a; LaVeck 2006a). If the cruelty frame can prioritize a discussion of how commercial interests dictate animal suffering and commodification in farming in almost *all* cases, even on smaller farms, then people may begin to see that there are not many farms or fishing practices that truly would be capable of eliminating animal suffering (or unfairness) and still turning a profit.

This frame of universal suffering in agriculture could, perhaps, be used for utilitarian purposes as a preliminary strategy to explain the reality of modern factory farming to the public and open the door to introducing the primary frame of injustice. Matt Ball stated that he viewed Vegan Outreach's strategy of focusing on factory farm suffering as a pragmatic "first step" (a foot in the door) for people who may then evolve toward animal rights over time. That can work if they are soon asked to make bigger changes after the first step (Cooney 2011), but I contend that these initial animal *welfare* frames must be supported to a greater extent by some *rights*-oriented frames, like injustice, if people are going to be overtly encouraged to begin to consider changing, not just their diet, but their exalted perceptions of *themselves* in relation to other animals. I believe a cruelty frame alone does not ideologically lead viewers toward a path of eventual transformation in deconstructing the human/animal dichotomy and challenging speciesism.

Another approach would be to reconceptualize the cruelty and suffering frame as a subcategory of an injustice frame, amplifying the idea of cruelty to not only mean causing someone to suffer *pain* but also to suffer the injustice of being enslaved and *used*. An example of this kind of message could be:

- "Animal use is animal abuse. Please eat veggie proteins."
- "Farming is (ab)using animals. Go vegan."
- "Don't make anyone suffer the indignity of being bred to be nothing but a snack."

Additionally, if cruelty frames continue to prioritize the most egregious cruelty and suffering, then factory farms will always get the most attention while fishing and less atrocious or intensive farms will seem unproblematic in comparison. Therefore, when suffering is problematized, it would be better to emphasize the suffering specifically involved in *death* (for both wild and domesticated "food" animal species) and in other basic agricultural practices that tend to be standard to *all* farms, including smaller, traditional farms. To further support the subject status of farmed animals, their suffering should be shown to include not only physical pain but also *emotional* pain due to family separation, frustration, boredom, social anxiety, and fear. To demonstrate that traditional animal farming has always involved suffering, even prior to the advent of modern factory farms, activists could cite Plutarch's description of the suffering of farmed animals several thousand years ago and the many nineteenth century descriptions of farmed animal suffering, particularly in slaughterhouses (Walters and Portmess 1999). We're not trying to return to "old MacDonald's farm," as that wasn't ever cruelty-free. We are promoting *animal-free* farming.

The challenge is finding visuals that could capture the notion of how *all* farmed and fished animals suffer and die to become food without always resorting to intensive factory farm or industrialized slaughterhouse images, which too severely limit the discourse to being anti-industrial. Visual examples could include:

- Images that show humans holding knives near farmed animals (or bloody butchers' aprons).
- Images of fish caught in fishing nets desperately gasping to breath, or with hooks in their cheeks (such as in PETA's anti-fishing brochures).
- Visuals of men artificially inseminating turkeys, cows or other animals, or females stuck in rape racks. This emphasizes how farmers forcibly breed animals for profit and sexually violate females in the process.
- A cow used for dairy being forcibly separated from her calf so humans can take her calf's milk.
- Images of suffering and death at hatcheries, stockyards, and slaughterhouses (Farm Sanctuary uses a lot of these) to demonstrate that suffering occurs before and after the ani-

mal's life on the "family" farm. (although, these do tend to *imply* industrial-only, unless otherwise clarified)

- I like the poignant image on the cover of Matthew Scully's (2002) book *Dominion,* as it simply shows a sheep pathetically and vulnerably laying with her feet all bound together by rope (presumably before slaughter or to be controlled in some way).

Environmental Destruction

The frame of environmental destruction should highlight that it is irresponsible to supply America's largely unnecessary demand for animal products, since animal-based diets require that all animals, particularly nonhumans, pay the cost for the waste and contamination of life-sustaining natural resources. This environmental frame takes the animal rights movement's goal of veganism and extends it out via ecological principles of interdependence to include the environmental movement's goals of preservation and ecological health. The chance to unify with the environmental protection movement on the issue of animal-based diets might create a wider appeal and more support via frame extension, which serves both utilitarian and deontological communication goals.

One of the main areas of overlap between animal ethics and environmental ethics is the mutual desire to protect wilderness areas and species from extinction, with animal rights privileging animal species as individuals and seeing plant species more as collective entities which are integral to maintaining the health of wildlife habitats (Regan 2002). While it is in the interest of both the animal and environmental protection movements to fight factory farming due to its excessive waste and pollution, it is also in the mutual interest of these movements to promote a plant-based diet as a sustainable solution (FAO 2006; Freeman 2010c; Singer and Mason 2006; Worldwatch Institute 2004). As animal rights organizations are dedicated to protecting the interests of *nonhuman* animals where they face discrimination and exploitation at the hands of humans, it seems appropriate for their food advocacy to also problematize an animal-based diet based on the basis that it unfairly disadvantages *wild animals* by unnecessarily threatening their lives and habitats. For example, FARM's greenyourdiet.org environmental site has a tab dedicated to "Wildlife" that talks about threats to fish as an ecologically-valuable species but also as individuals who feel pain. Animal activists should more overtly mention agribusiness/fishing's connection to extinction of species, loss or pollution of wild habitat (including damage caused by cattle grazing), and any government-sanctioned mass killing of "predatory" wild animals to protect rancher/fisher interests (such as the Canadian seal hunt and the USDA's Wildlife Services' trapping, poisoning, and shooting of millions of animals such as coyotes).

I am not suggesting that animal activists must ignore the negative environmental effects of an animal-based diet on *humans*, but it is more in keeping with the goals of animal rights to use its limited resources to speak out especially for *non*human animals wherever they are unfairly threatened by humans. This animal-centered environmental appeal addresses Erica Meier's concern that we are animal protection experts, not environmental experts. The environmental frame can also serve as a useful opportunity to deconstruct the human/animal dualism and promote the idea that humans are fellow animal beings (earthlings) who are dependent on the same ecosystems; therefore, humans should not selfishly take an excessive amount of the shared resources that *all* animals require for life.

As an example, environmental veg messages should not just be accompanied with the tagline "save the planet" but also "*share* the planet." A message declaring oneself a herbivore could give the rationale "to save animals (both wild and farmed)," or "to spare farmed animals and protect wild animals," or "to save animals and ecosystems."[3] And we could amend the environmental t-shirt slogan to say "Humans: We're not the only species on the planet. We just *eat* like it... Try Vegan."

Where to Place Blame for Problems

While a focus group study concluded that certain meat-eating consumers responded better to a pro-vegetarian frame that blamed agribusiness not them (Mika 2006), I posit that blame should be placed more on consumer demand and a speciesist society, as I argue for a

[3] Ideologically, I prefer the term "free" animal to "wild" animal (or "wildlife"), but I'm afraid that people may not understand whom I mean when I say "free animal." In text and conversation you can explain the distinction; on pithy bumper stickers and t-shirts, you can't.

more ideologically authentic approach than a utilitarian or strategic marketing approach. Rather than primarily blaming a dirty or greedy industry, blame should shift toward the simultaneous culpability of consumers for creating a *demand* for food products taken from animals and supporting injustice and killing, as this will better align with the problem frames I suggest in this chapter.

In chapter four I discussed the dilemma with making agricultural practices the problem, as it constrains the discourse to debating how animal agriculture or commercial fishing should *improve*. While it makes sense that if agribusiness and fisheries are to blame, then they should reform, this supports a *welfare* solution more so than a rights solution. Animal rights organizations should explain that, collectively, through America's legal system and, individually, through consumer choices, Americans personally and publicly support the exploitation of nonhuman animals for food and its resulting environmental destruction.

For example, in showing audiences a video of abusive behavior at a dairy farm, Gary Yourofsky (2010) doesn't just blame the abusive employee (although he admits that guy is a "scumbag"), but clarifies consumer culpability by noting that the employee is doing that because *we* want to eat what comes out of her body. And FARM has a poster of a cow whose neck is slit and bleeding with the phrase "It's a filthy business. They couldn't do it without you." And PETA's *Chew on This* DVD declares people should go vegetarian because: "it's violence you can stop," and "no animal deserves to die for your taste buds." Linking consumers to the problem (not just the solution) fits Derrida's (2004) projection that industrialized violence against animals will ultimately end when our self-image can no longer stand the spectacle of our own immoral behavior.

It is appropriate to acknowledge, as most animal organizations did, that consumers have not been given much information about the *extent of* the injustice, cruelty, or environmental destruction associated with animal-based foods, so part of the animal activists' job is to provide the public with information as evidence supporting these problem frames. Additionally, we should acknowledge that our society as a *whole* condones and naturalizes certain animal exploitation, requiring an interrogation of the Western worldview that unfairly dichotomizes humans from all animals. The prominence of carnism may help explain why individual citizens are generally willing to relinquish knowledge and awareness of exploitative practices and ignore issues facing farmed animals, as society encourages this compliance and lack of consideration (Adams 1990; Dunayer 2001; Joy 2010). As Derrida (2002) stated, "[men] do all they can in order to dissimulate this cruelty or to hide it from themselves, in order to organize on a global scale the forgetting or misunderstanding of this violence" (p. 394). This larger social validation for agricultural ignorance works to the advantage of each individual consumer who wants to eat animals with a clean conscience, fitting with the bias toward the status quo and social conformity tendencies in general (Cooney 2011). While I am tempering individual blame here in favor of putting it in a larger historical, socio-political context, once one's individual role in the system is elucidated, he/she bears a personal responsibility for creating solutions both as a consumer and a citizen.

While it is more challenging to place blame on the very public you are seeking to change, as it may be offensive, animal activists can experiment with different utilitarian rhetorical strategies to make the message less offensive, more empathetic, and more effective, as most of the activist messages are. For example, in terms of terminology, activists can more impersonally blame *society, carnism,* and *consumer demand,* or use *we* messages instead of more accusing and personal *you* messages. Or, alternatively, some instances and audiences may call for a refreshingly blunt approach, like lecturer Gary Yourofsky (2010) frustratingly declaring "enough already!" and Herbivore Clothing Company's message "Eat like you give a damn." Considering the scale of the mass slaughter, I think it does justify some straightforward expressions of exasperation at society's complicity in this travesty. I personally find it shamefully hypocritical that my culture brags about its values of freedom and justice while legally and financially supporting the captivity and killing of billions of individuals and refusing to mitigate or end the needless ab(use).

RECOMMENDED SOLUTION FRAMES

I recommend three main solution frames that relate to the suggested problem frames of injustice (due to exploitation and unnecessary killing) and environmental destruction (that harms wildlife, human animals, and habitats), as well as the lesser frame of cruelty and suffering inherent in farming and fishing. My three solutions address various aspects of an individual's identity and behavior:

(1) *You as a sentient animal:* Recognize and take pride in the mutual subject status of all animals as fellow sentient individuals; this includes not only respecting their/our right to life and liberty, but also showing compassion to avoid causing them to suffer,

(2) *You as a herbivorous animal:* Consume a satisfying, sensible, and sustainable plant-based diet to avoid exploitation, killing, and waste of resources, and work to create easier access to vegan foods in your community,

(3) *You as a political animal*: Work collectively as citizens to solve problems caused by fishing and animal agribusiness and construct a less speciesist society based on the notion of humanimal justice.

Collectively, these solutions address various aspects of our personal, socio-political, and biological identity as animal beings. The first is a values-based solution regarding one's perspective on his/her place in the world in relation to others, the second is a consumer-based behavioral solution regarding meeting the biological needs of oneself and others, and the third is a socio-political solution regarding working as an engaged citizen to democratically create a more just social system. In keeping with my thesis that a deontological communication strategy for animal rights must transform worldviews not just behaviors, a values-based transformation that redefines human identity is a critical part of the solution. And while a consumer behavior change is necessary and useful within a market economy, governmental and collective action solutions are also necessary, recognizing the target audience members as social citizens in addition to individual consumers. In the following section, I discuss each of these three solutions.

Values-Based Solution: Respecting the Mutual Subject Status of Fellow Sentient Animals

As most activist messages indicated, humans must begin to respect not just the ability of other animals to feel pain but also respect their mutual status as individual subjects of a life, or persons. To increase the relevancy and concreteness of this viewpoint, animal rights messages can continue to include common analogies between farmed animals and other nonhumans who are granted some subject status, such as dogs and cats. But to deconstruct the human/animal dualism, animal activist messages should also include analogies with the *human* animal and openly acknowledge that humans are also animals (embracing our humanimality). To reduce the humanism in this analogy, the frame should blend ideas of *kinship* based on evolution and sentience with ideas of *diversity* to celebrate that all animal species have unique traits that make them inherently valuable (Freeman 2010a). This helps avoid the suggestion that other animals have to emulate humans in all ways to be inherently valuable subjects. Directionally, we should emphasize how in many ways *we* are like *them*.

Then activists can ask Americans to take the values they hold for humans and other subjects and consistently apply them toward "food" animals, including fish, as well as toward the "wild" animals with whom they share the planet. This includes valuing *connection, life, freedom, justice,* and *respect*. A related subcategory is to acknowledge and encourage the popular *welfare/compassion* sentiment stating that we generally do not want to be responsible for causing suffering to other sentient beings, clarifying that farming and killing inherently involve some suffering.

While this values-based transformation is listed in the solution section, it does not mean that it must be listed so literally in the "what can you do" call-to-action message sections of advocacy communication. It may be used as part of the motivational component or to build a case for the problem component of the collective action frame, meaning these values should be embedded throughout all rhetorical appeals. I have listed the values/identity transformation here in the solution section mainly to reinforce its importance as a necessary component of the framing process in food advocacy so that solutions are not just defined as behavior-based.

Some examples that reinforce the individuality and interests of sentient animals and our humanimality include:

- PETA's collateral materials featuring an illustration of a chick declaring "I am not a nugget" and messages telling viewers that pigs and fish are "friends not food." (see images this page)
- PETA's *Chew on This* DVD messages that defend veganism from the nonhuman animals' perspective: "because no living creature wants to see her family slaughtered," "because when animals feel pain, they scream too," "because they feel fear," and "because they don't want to die."
- FARM's new poster and cards that ask us to "put yourself in their place," and "Make the Connection," juxtaposing pictures of *humans* next to farmed animals both enduring the frightening positions that agribusiness forces animals into: crowded in a cage, about to have the tip of one's nose/beak sliced off, being dragged to slaughter, and hung upside down by the legs on the slaughterline.[4] (see image next page)

[4] This bears some similarity to PETA's "Holocaust on Your Plate" vegetarian campaign juxtaposing factory farm victim images with Holocaust victim images, attempting to show that both types of oppression and mass killing are wrong. But those actual, historical images are likely perceived as more controversial than FARM's images (Freeman 2007). FARM's feature a variety of races, and attempt to have us empathize with agricultural injustice in a purely fictional "what if" scenario for humans.

- Use of unifying terms such as "nonhuman animals," "humans and other animals," "like them, we feel…" or "like us, they have the ability to…" as well as acknowledging what makes them unique/special/amazing as a species or as an individual.
- Use of gender-specific or personifying terms like *he, his, she, her, someone, somebody,* or *who*.
- Farm Sanctuary's videos and profiles that tell the story of individual, named animals and let us see and hear them communicating.

- Farm Sanctuary's new "Some*one* not Some*thing*" campaign that showcases scientific evidence to prove that the animal species we farm and fish have feelings and thoughts and therefore do not deserve to be treated as food objects. (farmsanctuary.org/learn/someone-not-something/)

Consumer-Based Solution:
Eating a Herbivorous/Plant-Based Diet

As all organizations stated, the premier solution to the problems caused by an animal-based diet is to eat a plant-based diet and avoid consuming products taken from animals. Eating vegan aligns as a logical behavior-based solution to my recommended problem frames of the injustice of animal farming/fishing's exploitation and killing of subjects as well as the resulting environmental destruction of wildlife and our shared natural resources.

Vegan messages should continue to include an appeal to the self-interested value of health, to a certain extent, as it is essential to the argument that killing animals is not necessary for humans to survive and thrive in most cases. Activists did this through educating the public about the health benefits of a plant-based diet and ways to avoid any health risks, as well as providing tips and tools for transitioning to veganism and maintaining the healthy diet for a lifetime. These health messages fit with vegetarian advocacy recommendations from the Humane Research Council (HRC) (2007). As an example, I like Farm Sanctuary's slogan "Veg for Life" and FARM's "Go Veg. It's good for your heart," as they both promote a self-interest in health in ways that also imply an altruistic support for saving lives and expressing kindness and love.

The activists' additional appeals to the value of having pleasurable and convenient food on a vegan diet serve a utilitarian purpose supporting the health frame, showing how the diet is feasible and fulfilling. Additionally, the symbolic use of the color green is useful and representative of a plant-based diet, both in terms of the diet's association with healthy, fresh green plants and with going "green" or environmentally-friendly living.

Promoting a plant-based diet, especially organic, is preferred to solutions that suggest consumers just reduce their consumption of animal products, as it is awkward for animal rights organizations to craft a moral message saying "why don't you exploit and kill *fewer* animals – just save whatever number of innocent individuals is convenient for you." My recommendation to promote veganism is more deontological than the utilitarian meat-reduction recommendation by the HRC (2007) in terms of what they think will be more resonant with the meat-eating public. Granted, it's easier to ask for less, so to be more strategic in asking for more, animal activists should avoid inflexible language implying veganism is a strict "all or nothing" proposition of dietary purity, as that phrasing is severe, competitive, and unrepresentative. The diet is simply about living up to your ideals by withdrawing your financial support from animal exploitation, which will help your physical and mental health, as well as the health of fellow earthlings. One unique way FARM's "Make the Connection" handout phrased the vegan call-to-action (without using the term *vegan*) was to highlight replacement, saying:

> The only way to end these atrocities is to replace meat, eggs, and dairy in your diet with the many ready-to-eat alternatives available in your local supermarket, as well as vegetables, fruits, nuts, and grains.

However activists find a way to make the vegan message appealing, ultimately I suggest they be true to their values in recommending that society transition away from using fellow animals.

Of course, some people will perceive that a request for veganism is asking for too drastic a lifestyle change, but it doesn't mean they will necessarily reject the message altogether and do nothing. On their own, they may go vegan later when they are ready or more fully convinced. Or they may give up meat now but continue eating eggs and dairy. Or they may switch to eating different animal species or reduce their overall meat consumption, as the HRC (2007) found a quarter of Americans are willing to do. It's generally true that every little bit helps and something is better than nothing, but let the conscious choice to still buy some animal products be the consumer's *own* idea and not the proposed compromise solution of the animal rights organization. Plenty of other organizations will settle for making the more "palatable" meat-reduction requests (such as health organizations, and, unfortunately, environmental organizations). Animal rights organizations need to raise the bar and brand veganism as a sensible, sustainable, and satisfying way of life.

Transitioning to an eco-friendly, cruelty-free diet needs both economic and social support. When people start local vegetarian societies or just organize "meet-ups" through social media, it helps keep individuals networked with other like-minded vegetarians and it mainstreams the diet so it feels more "normal" and socially accepted. Knowing that humans don't want to feel different or isolated, some organizations (like the Vegan Society in the UK) are assigning mentors to new vegans to help provide them with the social support and guidance needed to withstand the momentum of meat-based social norms. Organizing regular group dinner outings is another way to keep people socially connected while also showing restaurateurs that they can profit from offering vegan menu items.

To make it easier to access vegan foods, activists can encourage people to act locally to get more plant-based options in restaurants, grocery stores, hospitals, schools, and entertainment venues (like sports stadiums). This is especially necessary in low-income areas that might be "food deserts" where healthy produce or veggie proteins are not as readily available or affordable as fast-food animal-based proteins. To enable more neighborhoods to access plant-based foods, people can plant community gardens, get more farmer's markets in town, and host tables with free vegan food samples and coupons.

And in addition to asking existing restaurants to just add a portabella mushroom sandwich or coconut milk ice cream to the menu, I like to encourage them to "veganize" as many of their current offerings by substituting plant-based products for animal products as much as possible (thereby reducing the amount of animal products they are purchasing). These alternative ingredients are likely to be somewhat indistinguishable to the consumer in terms of taste, particularly in baked goods. For example, a coffee shop need not sell a vegan scone *and* also a "regular" scone (as if there are two different species of human, suggesting "regular" people would be harmed if they were denied the small amounts of egg and cow's milk in bakery items). As a default, there should just be one scone that is made as animal-free as it can be so that everyone can enjoy it.

In thinking about this last suggestion, I am struck with a potential contradiction it presents based on my recommendation to ask *individuals* to ultimately *eliminate* animal products from their diet but to ask *retailers/restaurants* to just *reduc*e animal products from their shelves/menus/ingredients. Why settle for meat reduction from institutions/businesses, yet expect veganism from individuals? For consistency why not ask both entities for the same thing – to eliminate animal products, or to just reduce animal product consumption, or to just add in more plant-based food choices?

The answer may depend on the *point* of getting more vegan options in restaurants/groceries/institutions. Is it more about making it easier for individuals to be vegan, thereby encouraging more people to adopt the lifestyle and eventually increase the overall demand from retailers? Or do we want retailers/institutions to offer vegan foods as the default in many cases so that more people eat and become accustomed to vegan foods, even if not consciously identifying with the cause? In both cases, the aggregate number of animals killed goes down, which is a goal. But a premise of this book's thesis is that it matters *how and why* we reduce the number of animals killed for human food, not only that we *do* reduce the number killed. It needs to primarily be a conscious, altruistic choice to spare animals so we identify ourselves with them as fellow animals and earthlings. Otherwise, if our socio-cultural values are not challenged and enhanced, many other types of animal and environmental exploitation (beyond the food issue) will remain unchallenged.

In thinking about that motivation, I suppose I am hoping that society will transition to a solely plant-based diet, first primarily starting with committed individuals who set the example as vegans and begin to create a critical mass, encouraging more from other people and institutions, where the latter entities may adjust to these trends and ideas gradually as they become economically and politically viable. Perhaps this higher expectation for individual integrity reflects that I am being ideological when it comes to people and pragmatic when it comes to institutions/businesses, knowing that the latter should be a reflection of the cultural values of the former (and not vice versa). Yet, a commercially-funded media system seeks to produce a consumer culture where individuals identify with the values and needs prioritized by sponsoring businesses and institutions (McChesney 1999). But when it comes to culture, the direction of influence doesn't just flow in one direction and various entities impact each other, so I'd like to see individuals seize their agency and inject more moral/ethical influence into media, the market, and government. To promote this cultural shift, animal activists should ask society to identify with and act primarily upon intrinsic values (Crompton and Kasser 2009), such as those I have listed in my chart.

While it is essential for individuals to withdraw financial support, where possible, from animal exploiting businesses, I do have a concern that the emphasis that activist messages place on the power of consumer choice can imply a neo-liberal viewpoint that social issues can and should be solved primarily through individual market choices rather than also through the accompaniment of legal/political reform, public education, and social movements. A "choice" frame also threatens to limit veganism to a consumer/lifestyle trend rather than an ethical obligation, political issue, and environmental necessity. However, messages asking people to "choose vegan" can better connect consumers with their role as citizens if accompanied by an appeal to the values of *moral integrity* and *desire to make a difference*, as that implies that each person's private actions have public consequences for which we should be accountable (acknowledging that "the personal is political").

Citizen-Based Solution: Working Collectively to Create a Just Humanimality

While animal activists favored the individual consumer solutions of changing one's diet, they sometimes more overtly engaged consumers as citizens, such as when Farm Sanctuary asked people to write legislators in favor of bills banning the killing of downed cows or the force-feeding of ducks, or when organizations engaged the public as potential activists who could join their organization and get involved in direct action or public education. In order to better enable a cultural transformation in support of veganism and animal rights, it is important that the consumer solution is not suggested in isolation of addressing the broader *systemic* issues in American culture, politics, and economics that support legalized animal exploitation and an animal-based diet over a solely plant-based one. To do so, animal organizations should engage their audience as citizens, asking them to take part in changing an exploitive system to protect the lives of other animals and support freedom over domestication. They should more actively try to provide a *vision* (Lakoff 2004) for the public of the kind of non-speciesist society Americans can create together

In considering legal/political action, animal rights activists should try to find collective action strategies that are in keeping with an animal *rights* philosophy instead of primarily suggesting "larger cage" reforms to the agricultural industry. As Alex Hershaft stated in our interview, he is not against welfare reforms coming from animal *welfare* groups, but he is against animal *rights* groups promoting it because it is not authentic to their anti-exploitation position:

> We are in favor of welfare reforms. We are just not in favor of animal rights advocating those because it leaves the wrong impression with the consuming public. It gives the impression that we approve of the use of animals - exploitation of animals - for food as long as they are treated a little less reprehensibly. We feel that welfare reform is something that the animal exploiting industry should be introducing to try and entice the consumers, the socially conscious consumers, to consume them.

Regarding long-term abolition, consider the Nonhuman Rights Project. While this legal effort isn't specific to animals used for human food, vegetarian advocates can be supportive of this legal campaign for nonhuman personhood. Headed by animal rights attorney Steven Wise, it seeks to establish legal personhood status for some key animal species under American common-law, so they would not continue to be categorized as things. This is a first step toward granting legal rights and standing to nonhuman animals in the United States legal system and could be adapted to fit each country's judicial system.

Other government/policy-based collective action solutions could include making animal agribusiness and commercial fishing industries accountable by paying for the environmental damage they cause or having their executives serve jail time for breaking environmental laws. For this to be effective, it would first require that citizens ensure that environmental laws do not exclude agriculture. If the animal agriculture industry had to internalize the costs it currently externalizes on society and other species, the price of animal products would likely rise, which might serve some utilitarian purpose of reducing overall consumption of animal products in America. Citizens could also demand that the government cease subsidizing animal-based agriculture (including plant crops used as farmed animal feed and low-cost cattle grazing on public lands) in favor of greater subsidies to *plant-based* agriculture, especially organic. And government facilities (such as prisons, military bases, and schools) could be encouraged to replace many animal-based foods with plant-based alternatives.

Citizens could demand more transparency from industry to publicly monitor its use of animals, such as requiring agribusiness to provide greater public and media access to all facilities, including slaughterhouses, with the ability to visually record routine practices. Imagine what would happen if we had live camera feeds streaming 24-7 on all USDA-regulated slaughterhouse killing lines.[5] There is also an issue with deception or lack of information in animal product labeling; Compassion Over Killing addresses this in their campaign for fair labeling on egg cartons. This requires more transparency from the animal agribusiness industry in labeling its products honestly for consumers regarding animal welfare conditions, feed and additives, GMO use, and environmental policies, raising the profile for those issues in the public's minds.

Related to this idea of increasing the transparency of the industry and the public awareness of the problems associated with an animal-based diet, activists could ask citizens to request that the news media

[5] If the undercover activist videos of farmed animal abuse that occasional make the news cause some consumers to reduce their meat intake (Tonsor and Olynk 2010), then having increased exposure to animal violence would theoretically motivate even greater dietary changes, while also potentially reducing some of the most egregious abusive behavior by employees who can't stand the spotlight. However, daily exposure to slaughter would run the risk of desensitizing people to animal killing, potentially further trivializing its significance.

put these topics on the agenda, not just from an anthropocentric public health and environmental standpoint but also from an animal rights standpoint that begins to question humans' right to use fellow animals in this way (Freeman, Bekoff, and Bexell 2011). And entertainment programming should also be encouraged to include perspectives and worldviews sympathetic to animal rights (see the HSUS's Genesis Awards for examples of pro-animal media). The music scene is also an important site of cultural production and social change messages, as some animal rights groups, such as FARM, Vegan Outreach, and PETA, are working at music festivals to support vegan bands. As media are no longer just top-down, citizens should be encouraged to produce their own media to post online that explores topics such as: vegan meals; problems and solutions proposed by the animal rights movement; and agricultural practices, specifically killing. Support for independent, non-commercial media is crucial to giving media producers the freedom to be critical of speciesist hegemonic structures.

Animal organizations could also recommend community-based collective action solutions, some of which the animal rights groups did in "get involved" sections online. For example, activists could suggest that adults and kids participate in schools, either through conducting humane education, starting student vegetarian societies, creating plant-based cafeteria options, and planting organic gardens. Around libraries or universities, people could screen documentaries on animal agriculture or hold public forums for discussion of humans' use of other animals for food. Citizens could be encouraged to participate in civil disobedience or public protests at local animal use facilities, such as at a slaughterhouse or hatchery. When feasible, people can adopt rescued farmed animals as companions or publicly support farmed animal sanctuaries to help provide more opportunities for humans to engage with these animals as fellow living beings instead of food objects.

When it comes to encouraging advocacy, I was impressed with how the animal rights organizations that I studied provided so much information on how people can take action and become grassroots activists. They not only sent out action alerts and provided the literature and materials that local activists could utilize, but they often trained people on how to be an effective advocate for animals. For example, Veganoutreach.org has many essays on what works and what doesn't and the philosophy behind advocacy. And Farm Sanctuary's Compassionate Communities campaign has a section on how to "be a better advocate." And every year FARM hosts a national animal rights conference to network and train activists.

In support of critical animal studies tenets, animal activists should also work in solidarity with other social movements to oppose all exploitation and oppression (such as for human groups facing discrimination) and seek fairer political and economic systems globally (Best et al. 2007). This integrated approach to promoting social justice would also mean ensuring campaigns on behalf of nonhuman dignity do not belittle the dignity of human groups seeking increased equality. I don't mean vegan campaigns should shy away from comparing human and nonhuman animal issues, rather that we ensure we are not helping one group of beings at the expense of the other. This coalition-building can have the added benefit of helping to diversify the animal and vegan movement – a needed improvement so the movement isn't seen as (or dismissed as) an elitist Western white movement (Best 2009).

My Recommendations Considered In Theory

This section provides a more collective analysis of my recommended frames, situating my ideas and findings within scholarly recommendations for advocacy communicators. This will enable more in-depth reflection on their strengths, weaknesses, and strategic value.

Ideological Authenticity with Expediency in Mind

In this book I often distinguish between deontological or "ideologically authentic" message strategies and more utilitarian strategies, yet tried to blend the best of both. Here I drew inspiration from the lessons of previous social movements, namely Campbell's (1989) analysis of ideological purity versus the more expedient approach in women's rights campaigns and Bormann's (1971) analysis of agitation versus conversionist approaches in the abolitionist movement. For example, expedient or conversionist strategies sometimes perpetuated or even used racist or sexist stereotypes to gain adherents rather than critiquing these prejudices as the source of the problem. Similarly, I argue that, rather than perpetuating speciesist values in an attempt to gain more widespread appeal that is often limited to behavioral changes and welfare reforms, animal rights activists must challenge the human/animal dualism and speciesist worldviews that serve as the basis for nonhuman animal exploitation.

Based on an analysis of factions within the abolition and civil rights movements, Bormann (1971) recommended that social movements stick to the strong moral values and rights rhetoric of the agitators to avoid watering down the message like the conversionists did. But conversely, activists should situate a rights message within national cultural values and heroic historic struggles, like the conversionists did, rather than using revolutionary or inflammatory rhetoric like the agitators. Animal rights organizations in this study loosely followed Bormann's advice by often using a moral message and having it be culturally resonant, positive, nonthreatening to the republic, and sometimes even patriotic. They also maintained credibility by not being misanthropic or advocating violence or hatred, as those values would be logically inconsistent within a moral movement (such as animal rights) built on increasing respect for life (Munro 1999; Singer 1990). However, animal activists were more expedient in their choice to moderate this moral rhetoric, constraining it to conversionist welfare appeals rather than explicitly appealing to more ideologically powerful concepts like rights and justice, as the agitators did.

Another expedient/conversionist tactic to be wary of is when, as part of the frame alignment process, animal rights appeals extend out to show their relevance to human-based causes that people care about (anthropocentric altruism) –such as how animal agribusiness is hurting humanity by contributing to world hunger, fostering unfair/unsafe work conditions, and polluting "our" land and water. While these can certainly be mentioned as yet another legitimate and important rationale for veganism (in solidarity with social justice struggles), animal activists should not resort to leveraging society's anthropocentrism as a tool to save nonhuman animals by default; that tactic does not challenge the human/animal dualism that is the root cause of animal exploitation and can serve to inadvertently reinforce the problematic attitude that human life is more inherently valuable than any nonhuman animal's life. That humanist attitude is why humanity thinks it is justifiable to breed, enslave, kill, and eat nonhuman animals in the first place.

The debates between purity/agitation and expediency/conversion strategies are similar to Cox's (2006) and Gitlin's (2003) acknowledgement that a key communication dilemma social movements face is balancing how critical they can be while still maintaining credibility by staying within "symbolic legitimacy boundaries" (Cox 2006, 61). Gitlin (2003) noted that social movements walk a line between being assimilated and "blunted" (290) if they are too moderate and being marginalized and trivialized if they are too critical. In this study, I concluded that animal rights organizations' common use of animal *welfare* values is moderate enough to gain them legitimacy, yet the sheer magnitude of the animal suffering they expose on factory farms is powerful enough to keep even this moderate welfare message from being blunted. However, this suffering frame runs the risk that industry can counter-frame itself as solving the problem through "humane" reforms, however misleading that may be, thereby assimilating the issue and becoming animal welfare proponents themselves. Therefore, I propose that animal rights activists should frame killing and exploitation as the problem so their messages retain a critical and ideologically-authentic edge, befitting Ryan's (1991) recommendations. While this more critical frame certainly does run the risk of the message being marginalized, the risk is reduced if activists skillfully use frame transformation alignment around the meta-frame of justice.

Increasing the Resonance of Transformative Frames

Frame transformation alignment can be facilitated by using a broad or global interpretive frame, such as a meta-narrative, which reframes many domains of life under a new universe of discourse (Snow et al. 1986). I propose that *justice* be the global interpretive frame that animal rights activists should use to create frame transformation. To do so, activists first need to engage a more direct comparison of the sentience and individuality of farmed animals to the *human* animal so that humans will be challenged to recognize their own humanimal status and the farmed animal's own status as a fellow subject of a life. This alignment process would then articulate that, for moral consistency and fairness, many of the major justice values Americans already hold in favor of protecting humans and their rights, such as *compassion, respect, life, fairness,* and *freedom*, should transfer to protecting other animal subjects. This is the kind of society I envision when I say a "just humanimality." These two major, related transformation frames can be summarized as stating that we are all animals, and, therefore, we should all have the same basic rights to "life, liberty, and the pursuit of happiness" (a phrase drawn from American political rhetoric). The values are abstract enough to fit within rhetoricians' recommendations for creating widespread support based on appealing to culturally-accepted principles that are powerful in large part because of their ambiguity (McGee 1980; Perelman and Olbrechts-Tyteca 1969; Therborn 1980).

As examples of campaigns that connect human and animal rights, the Vegan Society in the UK created a short introductory video for their vegan pledge campaign, entitled "Do you want to make history?" It compares Western civilization's progress toward social justice for humans to a current need to do the same for "fellow animals" whom we kill for food. The video implicitly and explicitly highlights many of the justice values that I recommend (*life, freedom, respect, integrity; kinship, community/connection; empowerment, growth, responsibility, importance/usefulness,* and *democracy*), when the vegan call-to-action declares:

> If you believe all our fellow animals seek life and freedom, imagine being strong enough to follow your own convictions. Yesterday they made history [Gandhi, Martin Luther King, Jr., and Rosa Parks] today we are choosing to respect all animals. Live vegan. For a week…for a month…for life. Make History. VeganPledge.com.

It never relies on any abusive footage nor does it mention factory farming (it just uses single images/portraits of animals – human and nonhuman). It simply addresses the disconnect between people's beliefs and their actions, asking them to have the courage to take the logical next step in society's moral development. Another good example is PETA's 24-page general animal rights booklet entitled "All Animals are Equal" that explains "why supporters of liberation for women and minorities should support animal liberation too" by sharing a lengthy excerpt from *Animal Liberation* (Singer 1990) and highlighting various animal exploitation practices, including eating animals. It is unique for an animal rights organization's brochure to actually mention animal rights, explain its philosophical basis, and put it in context of human rights. This fits with critical animal studies tenets to see animal rights as part of a larger social struggle against all forms of oppression (Best 2009; Torres 2008; Nibert 2013). Embedding veganism within the bigger picture of social and ecological justice encourages people to base consumption decisions not just on a narrow version of vegan purity but also on other justice-oriented factors that facilitate fairness toward humans (non-sweatshop or fair trade items) and nature (organic or recycled items) (Torres 2008).

In addition to a human rights comparison, another case of frame transformation in support of animal rights ideology would be a frame comparing the rights of nonhuman animal species whom we have *domesticated* to live as freely and naturally as *wild* nonhuman species do. This animal rights ethic would also loosely align with a deep ecology ethic that values the naturalness and freedom of wild animals to live less hindered by excessive or unnecessary human interference as fellow animal species who contribute to the health of the ecosystem (Devall and Sessions 1985). To consider the rights of historically domesticated animals not to be domesticated and exploited, especially when unnecessary for human survival, seems like a radical transformation in American worldviews, which would qualify it as a frame transformation in my estimation. Additionally, a focus on problematizing killing and captivity as an injustice against freedom would significantly work toward a long-term strategy that challenges an instrumental and anthropocentric worldview, something especially pertinent to movements working on behalf of other species (Evernden 1986).

In openly stating their vision that no animal should be domesticated, activists could point to the rationale that agricultural domestication is largely uncommon according to nature's principles and morally illogical according to human society's principles (Nibert 2013). This vision would be drawing upon principles from both human rights *and* environmental ethics to request that our relationship with nonhuman animals become both more respectful and natural. This encourages a *blend* of natural and cultural ethics principles in governing how humans treat nonhumans (Freeman 2010a; Jasper and Nelkin 1992; Pollan 2006). This would also support some scholars' desires for more unification between the animal and environmental protection movements (Beers 2006; Freeman 2010c; L. Hall 2006a; Jasper and Nelkin 1992; Maurer 2002). And it does so using an ideologically authentic frame transformation process that *directly* supports animal liberation principles instead of solely using a frame extension process that encourages people to stop eating animals for environmental or self-preservation reasons.

A challenge for activists is how to make these abstract master frames of justice, rights, or sustainability relate to each audience member's personal experiences by not seeming too distant from his or her everyday life (Benford and Snow 2000). I think the animal rights organizations connected with people on a personal level by emphasizing how going vegan allows you to make a difference daily and have some control "at every meal" or "with every bite." Additionally, they tried to connect with Americans' personal lives by alluding to their beloved companion animals and comparing how they were similar in personality, intelligence, and emotions to the animals we farm/eat. Asking people to envision their own bowl of raisin bran or their own dog Sam helps make it real for the audience, so they can better connect with these otherwise overwhelming global issues.

Changing Worldviews

Another challenge social movements face is changing entrenched worldviews – attempting to reveal that the public's accepted view of reality is based on a faulty premise (Walters and Portmess 1999). To reveal the faulty premise behind the American public's acceptance of carnism, animal rights organizations provided evidence that (1) farmed animals endure much suffering to become/produce our food items and (2) these resulting food items are not required for a healthy diet. These are key assumptions to challenge so people can't continue to assume that: (1) we are the most moral omnivore because we raise and kill our prey in a supposedly civilized, humane, and nonviolent way, and (2) animal farming, while it does involve some violence, is simply a necessary evil to sustain human life.

I believe an even more fundamental faulty premise to address is the "human moral superiority complex" – that humans, as some kind of pinnacle or divinely "chosen" species, are entitled and justified in viewing other animals as a resource. The new premise should be that each sentient being is a fellow individual whose life and freedom matters and deserves the respect we give our fellow human beings. If the sanctity of animal life became a new assumption, it would challenge the premise that it is ethical for humans to kill fellow animals when it is not in self-defense, such as for research, entertainment, and unnecessary food items. This necessity argument was a basis for denouncing meat-eating as immoral even as far back as Ancient Greece, by vegetarian scholars such as Pythagoras (Walters and Portmess 1999). Therefore I recommend that animal activists problematize killing and farming anyone as unfair practices based on reconsidering who counts as "someone." If we see ourselves as similar to our fellow animals, then animal farming does not simply become inhumane, it becomes unjustifiable.

Demonstrating Flexibility to Avoid Extremism

If the public came to accept that animal farming and fishing is a problem because it is unjust, unnecessary, and environmentally destructive, then a natural solution would be shifting to a plant-based diet. Rather than suggesting that people merely cut back on their animal product consumption, animal rights activists should be clear in promoting a rejection of animal agriculture. Benford & Snow (2000) would rightly contend that such a boycott is too rigid and exclusive, thereby reducing the solution's appeal to a larger number of adherents. But I believe it demonstrates increased logical consistency and credibility for animal rights advocates to adhere closer to their own principles that prohibit the exploitation of fellow animals as an unnecessary food resource. To increase the flexibility of the vegan solution frame, the animal activists can continue their practice of mentioning that switching from animal to plant-based foods involves a *transition period*. They expressed understanding that going vegan is an experimental and sometimes slow process that requires adaptation and doesn't happen overnight.

Promoting similar theories about the benefits of ideas being elastic, Perelman and Olbrechts-Tyteca (1969) suggested that communicators make their notions flexible, adaptable, and progressive while making their opponents' ideas seem rigid and outdated. Toward this end, many animal organizations did highlight flexibility by discussing the *ease* of consumers choosing *plentiful* vegan products. And moral progressiveness was implied by placing veganism within a *moral integrity* frame, based on compassion or environmental responsibility.

And conversely, activists portrayed opponents as rigid in the sense that industry allowed profit motives to blind them when it came to their obligations about animal welfare and environmental stewardship. Regarding the recommendation that opponents be characterized as out-of-touch, animal rights organizations tended to malign factory farming not for being *outdated* but for being *too modern*, huge, technological, exploitative, and destructive in opposition to bucolic ideals of more traditional American family farming. The problem with this characterization of farming as too industrial/modern is that the flexible and reasonable middle ground then becomes eating fewer animal products, but ensuring they come from so-called "humane" traditional farms or from the wild, which aligns with Michael Pollan's (2006) idea of a conscientious omnivore. But psychologist Melanie Joy (2011) declared that even conscientious omnivores are rigid in maintaining a stance that insists on arbitrarily seeing certain species of animals as edible and others as morally off limits to consumption. Vegans demonstrate more openness and flexibility by expanding their circle of compassion to include other animal species.

The challenge for animal rights organizations, then, is to appear flexible while still remaining firm in their ethical stance advocating for a vegan diet and the right of animals not to be farmed and domesticated. I think the diet can be shown to be flexible and less restrictive when activists highlight the *variety* and *plentitude* of plant-based food options that people can (and perhaps already do) enjoy. This includes showing how there are many alternatives to animal flesh and animal milk (soy, seitan, coconut, almond, pea, oat, rice, hemp, etc.) that have similar taste and texture and provide vital protein and nutrients.

I propose increasing the moral flexibility of vegan appeals by incorporating ecological or natural principles of predation (hunting) into our cultural principles governing whom we can/should kill for food. We can acknowledge that hunting of wildlife may be necessary in limited survival circumstances, as omnivorous or carnivorous animals hunt. We do so while declaring that human cultural principles of justice and rights (when used to govern human behavior toward fellow animal subjects) dictates that killing is only justified when done in self-defense or in times of extreme necessity. It is important that the "opponent" not be limited to just factory farming but that *animal agriculture* itself be shown as outdated, not technologically, but according to progressive *morals* that acknowledge the subject status of fellow animals and condemn the slavery, exploitation, and unnecessary killing of other subjects of a life.

Rigidity can also be a problem if activists create an identity that is too narrow or elitist, something Tarrow (1998) warned against. So rather than casting vegans as an elite subculture, I think animal activists did a good job of building a broader vegan identity around popular values such as desire to make a difference, integrity, compassion, and respect for life. This befits Crompton and Kasser's (2009) recommendations that all institutions and campaigns promote intrinsic social values. In this way, veganism was framed as politically and morally significant enough to create a broad-based positive identity for a vegan as someone who was altruistic and responsible, without limiting it to a certain demographic or cultural style – meaning they didn't characterize vegans as punks, hippies, females, urbanites, or models.

But being too moral can itself be deemed rigid, subjecting the "do-gooder" to potentially being ridiculed as pompous, self-righteous, comical, or hypocritical. To temper this tendency with ethical veganism, I like Kathy Freston's (2011) reasonable notion that veganism is not about perfection but progress. Perhaps no one can achieve the mythical "level 5 vegan" status satirized in *The Simpsons*.[6] To remind us that there is no such thing as a perfectly cruelty-free diet, vegan author Mark Braunstein (2010) proclaimed that "vegetarians are not a better sort of people, just a better sort of carnivore; and carnivores are just a better sort of cannibal" (17). This intriguing declaration acknowledges that harmlessness comes in degrees, and that not even a vegan is innocent of any killing (as we indirectly allow the killing of insects or unknown wildlife in cropfields), and even meat-eaters have some discrimination on whom they won't kill for food, namely their own species. As much as I promote framing based on moral values, I have come to realize that there is no ethical stance that is universally perfect or without contradiction at some level. If extended out to extremes, every principled practice has its flaws. I think this is what Braunstein meant when he provocatively stated, "any argument for vegetarianism can be extended as well for starvation, but then so could any reasons for carnivorism be extended for cannibalism" (137). While this is a wise statement about avoiding extremes, it doesn't dissuade Braunstein or me from recommending that by buying vegan we try to reduce harm to the extent it is feasible and beneficial. Offering similar encouragement to do as much as we can, even though we know we can't be completely harmless, Gary Yourofsky (2010) asked

---

6 This line is from a *Simpsons* episode entitled "Lisa the Treehugger" (Selman 2000) where an environmentalist demonstrates annoying ethical one-ups-manship, telling Lisa that her vegetarianism is a start, but, as a level five vegan, he (impossibly) doesn't eat anything that casts a shadow.

"Is there a reason we have to maximize the suffering that they already endure at our hands by also eating them?" He characterized our current diet as "radically cruel" and a vegan diet as "radically kind."

Pragmatically acknowledging that purity isn't possible, Vegan Outreach literature does a good job of reminding us veganism isn't about avoiding every trace of animal-based ingredients from a tedious long list. It's simply about making daily food choices that reduce as much suffering as you can. The focus should be on what veganism provides (mentally, physically, and environmentally) not what it lacks. This fits with Cooney's (2011) and Lakoff's (2004) suggestions to portray positive associations with your ideas/terminology, rather than negating and thus repeating myths and accusations leveled against you. As examples of phrases that associate veganism only with *positive* notions (in the affirmative), consider using:

- Go veg for life [Farm Sanctuary's campaign] (see image)
- Veggie meat saves lives; Veggie meat – real food for real people; Veggie meat – cool food, cool planet
- Compassionate choices [Vegan Outreach booklet title]
- I'm a friendly animal, eating animal-friendly foods
- Fair and fulfilling foods for everyone
- Environmentalism, Feminism, Veganism; Real feminists eat soy ice cream; Civil Rights, Civil Foods. Eat Vegan
- Respect animal dignity: eat animal-free foods
- Veganism for peace (of mind)
- Veganism: life, liberty, and the pursuit of happiness for all animals
- Eat more veggies – save the world [t-shirt slogan from A Well-Fed World]
- Serving a cause by serving vegan food; Sharing the planet by sharing vegan meals
- The world needs you to step up to the veggie plate
- Join the vegan movement; Compassionate communities [Farm Sanctuary campaign]; Team vegan
- Vegan vitality; Plant strong [Associated with the *Engine 2 Diet* (Esselstyn 2009)]; Healthy happy herbivore
- Vegan by nature
- Veganism is lean, clean, and green.

As examples of phrases that associate animal meat/farming with *negative* notions, consider using:

- Kick the meat habit [this is FARM's meatout slogan]
- Boycott animal exploitation industries; Just say no to farming anyone
- Farming / Fishing is a deadly business; Eating meat (ab)uses animals; Animal-abusing diet
- Avoid stealing what belongs to someone else – animals' milk, eggs, and bodies

- Don't hire a hitman just to eat lunch – meat is murder on animals
- (to describe free-range/family farming) less atrocious farming; less cruel farming
- Meat-eating is mean; Dairy and eggs are sexist; Flesh, eggs, and dairy are oppressing and depressing.

Another example is Farm Sanctuary's brochure "Factory Farming: Destroying the Environment," as it uses uniformly negative section headings: compromised air, deforestation and desertification, wasted water, fecal contamination, dead zones, ransacked oceans, and ruin on the range.

Inspiration to Act

As people will prioritize values differently, Perelman & Olbrechts-Tyteca (1969) suggest that rights activists help audiences rank values by emphasizing *quality* over quantity. Quality focuses on the rightness or uniqueness of concepts or individuals, as that which is threatened, irreparable, or priceless is deemed more valuable than that which has popularity (quantity) in its favor. This aligns with my suggestion that vegan advocates emphasize respect for the sentience, individuality, and life of fellow subjects as inherently valuable as well as continuing to emphasize morality and altruism more so than self-interest. Individually profiling the personalities, interests, and stories of rescued farmed animals can help life seem precious and unique, as stories help people connect to the larger issue in a memorable way (Polletta 2006). The environmental sustainability frame is also relevant to a quality-based appeal for veganism, especially regarding concerns about saving endangered species and preventing irreparable damage such as climate change and deforestation.

In addition to appealing to qualities that people value, activists must inspire agency to act by assuring the public that their assistance will indeed result in a better future and that overcoming the status quo is possible with their help (Snow et al. 1986; Stewart, Smith, and Denton 1984). I found that the animal rights organizations' messages were positive and encouraging about how each person's vegetarianism makes a difference and saves lives. But, considering the vastness of the problem and its roots in human history for thousands of years, the animal activists are challenged to provide a vision for a future without animal farming and exploitation. Instead, they tend to focus on the power of each individual to do the right thing and personally feel good about his/her choices. But emphasizing a solution frame also based on promoting collective action by engaged citizens can help toward creating a vision of how particular acts of incremental abolition will eventually lead to more systemic abolition of nonhuman exploitation. This fits with Torres (2008) recommendation that activism not just be framed as the domain of professional organizations (to which individuals just contribute money) but is primarily a personal responsibility, cultivated communally.

A Caveat: Keeping Independence In Mind

In speaking to these experienced animal rights leaders about their strategic communication strategies, I became concerned that I was suggesting they make strategic changes that might not fit with their more pragmatic goals as non-profit organizations (some of whom are relatively small organizations). While they are animal *rights* organizations, they may not share my belief that their organization must or should promote a critical animal rights discourse that seeks a change in worldview along with, or prioritized above, a more tangible improvement in short-term behavior. Therefore, in advocating my thesis, I may run into the problem that most activists actually promote *campaign* rhetoric which is more moderate than *critical* rhetoric (Cox 2006). If this is the case, it is not certain *whom* is supposed to promote the critical rhetoric of animal rights if it is not the leading national organizations within the animal rights movement.

Perhaps it might just be scholars, authors, and independent/grassroots activists, as they are freer to speak candidly than are organizations representing members, and they are less burdened by fundraising concerns that necessitate that they achieve tangible progress and short-term victories (Torres 2008). But, paradoxically, independent activists/scholars may lack the resources or organized networks to adequately mass communicate their critical rhetoric. Perhaps this is why the Humane Research Council (2007) pragmatically recommended that the animal protection movement as a whole should employ a *variety* of appeals, both critical and moderate, from different organizations. It also reminds us that each animal rights organization is a unique entity; and the organizations I profiled in this book vary in size and history, which affects their ability to mobilize resources and capitalize on political opportunities – factors that were beyond the scope of this book's analysis.

For the discourse of animal rights organizations to pose a critical challenge to speciesist worldviews, activists need not always directly promote animal rights *philosophy*, per se, using academic references and terminology. But I propose that whatever frames animal organizations do choose should be supportive of and informed by animal rights ideology instead of animal welfare or humanism so that they are logically aligned to pose a needed philosophical challenge to the root cause of exploitation, the human/animal dualism. If a convincing case has been built toward this thesis, then animal rights activists globally can hopefully find some ideas in the framing recommendations of this book that they could apply in their message construction and still be in keeping with the strategic approach that they determine to be successful based on their own experienced perspectives gained through decades of working on the worthy goal of ending injustice toward our fellow animals.

Works Cited

About COK. 2012. "About Compassion Over Killing." http://www.cok.net/about/.

About FARM. 2012. "About FARM (Farm Animal Reform Movement)." http://www.farmusa.org/about.htm.

About Farm Sanctuary. 2012. "About Us." http://www.farmsanctuary.org/about/.

About PETA. 2012. "All About PETA." http://www.peta.org/about/learn-about-peta/default.aspx.

About Vegan Outreach. 2012. "About Vegan Outreach." http://www.veganoutreach.org/about/index.html.

Academy of Nutrition & Dietetics. 2009. "Position Paper on Vegetarian Diets." http://www.eatright.org/about/content.aspx?id=8357.

Adams, Carol J. 1990. *The Sexual Politics of Meat: A Feminist-vegetarian Critical Theory*. New York: Continuum.

———. 1995. "PETA and a Pornographic Culture: A Feminist Analysis of 'I'd Rather Go Naked Than Wear Fur'." *Feminists for Animal Rights* 8 (3-4): 1, 8.

———. 2003. *The Pornography of Meat*. New York: Continuum.

Baker, Sherry, and David L. Martinson. 2001. "The TARES Test: Five Principles for Ethical Persuasion." *Journal of Mass Media Ethics* 16 (2-3): 148–175.

Balcombe, Jonathan P. 2006. *Pleasurable Kingdom: Animals and the Nature of Feeling Good*. London; New York: Macmillan.

Ball, Matt, and Bruce Friedrich. 2009. *The Animal Activist's Handbook: Maximizing Our Positive Impact in Today's World*. New York: Lantern Books.

Barclay, Eliza. 2012. "A Nation of Meat Eaters: See How It All Adds Up." *National Public Radio*. http://www.npr.org/blogs/thesalt/2012/06/27/155527365/visualizing-a-nation-of-meat-eaters.

Bauston, Gene. 2006. "It's Not a Black and White Issue." *Satya*, Sept.

Beers, Diane L. 2006. *For the Prevention of Cruelty: The History and Legacy of Animal Rights Activism in the United States*. Athens, Ohio: Ohio University Press.

Bekoff, Marc, and Jessica Pierce. 2009. *Wild Justice: The Moral Lives of Animals*. Chicago: University of Chicago Press.

Benford, Robert D. 1987. *Framing Activity, Meaning, and Social Movement Participation: The Nuclear Disarmament Movement.* University of Texas at Austin.

Benford, Robert D., and David A. Snow. 2000. "Framing Processes and Social Movements: An Overview and Assessment." *Annual Review of Sociology* 26 (January 1): 611–639.

Berbrier, Mitch. 1998. "'Half the Battle': Cultural Resonance, Framing Processes, & Ethnic Affectations in Contemporary White Separatist Rhetoric." *Social Problems* 45 (4) (November 1): 431–450.

Best, Steven. 2009. "The Rise of Critical Animal Studies: Putting Theory into Action and Animal Liberation into Higher Education." *Journal for Critical Animal Studies* 7 (1): 9–52.

Best, Steven, Anthony J. Nocella, Richard Kahn, Carol Gigliotti, and Lisa Kemmerer. 2007. "Introducing Critical Animal Studies." *Journal for Critical Animal Studies* 5 (1): 4–5.

Black, Jay. 2001. "Semantics and Ethics of Propaganda." *Journal of Mass Media Ethics* 16 (2/3) (June): 121–137.

Bormann, Ernest. 1971. *Forerunners of Black Power: The Rhetoric of Abolition.* Englewood Cliffs, N.J.: Prentice-Hall.

Braunstein, Mark Mathew. 2010. *Radical Vegetarianism: A Dialectic of Diet and Ethic.* New York: Lantern Books.

Brown, Kenneth L. 2004. "Ethnographic Analogy, Archaeology, and the African Diaspora: Perspectives from a Tenant Community." *Historical Archaeology* 38 (1) (January 1): 79–89.

Callicott, J. Baird. 1993. "The Conceptual Foundations of the Land Ethic." In *Environmental Philosophy: From Animal Rights to Radical Ecology*, edited by Michael E Zimmerman, 110–134. Englewood Cliffs, NJ: Prentice-Hall.

Campbell, Karlyn Kohrs. 1989. *Man Cannot Speak for Her*. Vol. One. New York: Greenwood Press.

Carey, James W. 1989. "A Cultural Approach to Communication." In *Communication as Culture: Essays on Media and Society*, edited by James W. Carey, 13–35. Boston: Unwin Hyman.

Cavalieri, Paola. 2006. "The Animal Debate: A Reexamination." In *In Defense of Animals: The Second Wave*, edited by Peter Singer, 54–68. Malden, MA: Blackwell Pub.

Chandler, Daniel. 2002. *Semiotics: The Basics*. New York: Routledge.

Chilton, Paul, Tom Crompton, Tim Kasser, Greg Maio, and Alex Nolan. 2012. "Communicating Bigger-than-self Problems to Extrinsically-oriented Audiences." UK: Common Cause Research. www.valuesandframes.org.

Clark, Jonathan L. 2012. "Ecological Biopower, Environmental Violence Against Animals, and the 'Greening' of the Factory Farm." *Journal for Critical Animal Studies* 10 (4): 109–129.

Clark, Stephen R. L. 1994. "Apes and the Idea of Kindred." In *The Great Ape Project: Equality Beyond Humanity*, edited by Paola Cavalieri and Peter Singer, 113–125. New York: St. Martin's Press.

Cohen, Joel. 2010. "Meat". First Annual Malthus Lecture, sponsored by The Population Reference Bureau & The International Food Policy Research Institute, March 3, Washington, D.C. www.prb.org/pdf11/cohen-lecture.pdf.

COK Financial. 2012. "Guidestar: Nonprofit Report. Annual Revenue 2012." http://www.guidestar.org/organizations/52-2034417/compassion-over-killing.aspx.

COK.net. 2012. "Compassion Over Killing Homepage." http://www.cok.net/.

Cooney, Nick. 2012. "Welfare Reform and Vegan Advocacy: The Facts, Farm Sanctuary – Compassionate Communities." http://ccc.farmsanctuary.org/welfare-reform-and-vegan-advocacy-the-facts/.

———. 2011. *Change of Heart: What Psychology Can Teach Us About Spreading Social Change*. New York: Lantern Books.

Cox, Robert. 2006. *Environmental Communication and the Public Sphere*. Thousand Oaks, CA: Sage Publications.

Crompton, Tom. 2008. "Weathercocks and Signposts: The Environmental Movement at a Crossroads." World Wildlife Fund-UK, Strategies for Change project. wwf.org.uk/strategiesforchange.

Crompton, Tom, and Tim Kasser. 2009. *Meeting Environmental Challenges: The Role of Human Identity*. Surrey: WWF-UK.

Cudworth, Erika. 2008. "'Most Farmers Prefer Blondes': The Dynamics of Anthroparchy in Animals Becoming Meat." *Journal for Critical Animal Studies* 6 (1): 32–45.

Davis, Steven L. 2003. "The Least Harm Principle May Require That Humans Consume a Diet Containing Large Herbivores, Not a Vegan Diet." *Journal of Agricultural and Environmental Ethics* 16 (4): 387–394.

Deckha, Maneesha. 2008. “Disturbing Images: PETA and the Feminist Ethics of Animal Advocacy.” *Ethics & the Environment* 13 (2): 35–76. a9h.

DeLuca, Kevin. 1999. *Image Politics: The New Rhetoric of Environmental Activism*. New York: Guilford Press.

DeLuca, Kevin, and Jennifer Peeples. 2002. “From Public Sphere to Public Screen: Democracy, Activism, and the ‘Violence’ of Seattle.” *Critical Studies in Media Communication* 19 (2) (June): 125.

Derrida, Jacques. 1976. *Of Grammatology*. Translated by Gayatri Chakravorty Spivak. Baltimore: Johns Hopkins University Press.

———. 2002. “The Animal That Therefore I Am (More to Follow).” Translated by David Wills. *Critical Inquiry* 28 (2): 369.

Derrida, Jacques, and Elisabeth Roudinesco. 2004. *For What Tomorrow: A Dialogue*. Translated by Jeff Fort. Stanford, CA: Stanford University Press.

Devall, Bill, and George Sessions. 1985. *Deep Ecology: Living as if Nature Mattered*. Salt Lake City, Utah: Peregrine Smith Books.

Donovan, Josephine, and Carol J. Adams. 2007. *The Feminist Care Tradition in Animal Ethics: A Reader*. New York: Columbia University Press.

Dunayer, Joan. 2001. *Animal Equality: Language and Liberation*. Derwood, Md.: Ryce Pub.

———. 2006. “Serving Abuse.” *Satya*, October.

Eley, Geoffrey. 1992. “Nations, Publics, and Political Cultures: Placing Habermas in the Nineteenth Century.” In *Habermas and the Public Sphere*, edited by Craig J Calhoun, 289–339. Cambridge, MA: MIT Press.

Elstein, Daniel. 2003. “Species as Social Construction: Is Species Morally Relevant?” *Journal for Critical Animal Studies* 1 (1).

Entman, Robert M. 1993. “Framing: Toward Clarification of a Fractured Paradigm.” *Journal of Communication* 43 (4): 51–58.

Esselstyn, Rip. 2009. *The Engine 2 Diet: The Texas Firefighter's 28-day Save-your-life Plan That Lowers Cholesterol and Burns Away the Pounds*. New York: Wellness Central.

Evernden, Neil. 1986. *The Natural Alien: Humankind and the Environment*. Toronto: University of Toronto Press.

FAO. 2006. “Livestock a Major Threat to Environment.” *Food and Agriculture Organization of the United Nations*. November 29. http://www.fao.org/newsroom/en/news/2006/1000448/index.html.

FARM Financial. 2012. "Guidestar Nonprofit Report for FARM." http://www.guidestar.org/organizations/52-1302627/farm-animal-reform-movement.aspx.

Farm Sanctuary Financial. 2012. "Farm Sanctuary 2011 Annual Report." http://www.farmsanctuary.org/about/financial/.

Farm Sanctuary Positions. 2012. "Position Statements: 'Humane Meat,' Welfare Reforms, In Vitro Meat." http://www.farmsanctuary.org/about/position/.

Farm Sanctuary Victories. 2012. "Victories and Milestones." http://www.farmsanctuary.org/about/milestones/.

Farmsanctuary.org. 2012. "Farm Sanctuary Homepage." http://www.farmsanctuary.org/.

Finsen, Lawrence, and Susan Finsen. 1994. *The Animal Rights Movement in America: From Compassion to Respect*. New York: Twayne.

Foer, Jonathan Safran. 2009. *Eating Animals*. New York: Little, Brown & Company.

Foucault, Michel. 1980. *Power/Knowledge*. Brighton, Sussex: The Harvester Press.

———. 1990a. "The Archaeology of Knowledge." In *The Rhetorical Tradition: Readings from Classical Times to the Present*, edited by Patricia Bizzell and Bruce Herzberg, 1130 – 1154. Boston: Bedford Books.

———. 1990b. "The Order of Discourse." In *The Rhetorical Tradition: Readings from Classical Times to the Present*, edited by Patricia Bizzell and Bruce Herzberg, 1154 – 1164. Boston: Bedford Books.

———. 2000. "So Is It Important to Think?" In *The Essential Works of Michel Foucault, 1954-1984*, edited by Paul Rabinow and James D Faubion, 3:454–458. New York: New Press.

Fox, Michael Allen. 1999. *Deep Vegetarianism*. Philadelphia, PA: Temple University Press.

Francione, Gary L. 1996. *Rain Without Thunder: The Ideology of the Animal Rights Movement*. Philadelphia, PA: Temple University Press.

Francione, Gary L., and Robert Garner. 2010. *The Animal Rights Debate: Abolition or Regulation?* New York: Columbia University Press.

Freeman, Carrie P. 2007. "Who's Harming Whom? A PR Ethical Case Study of PETA's Holocaust on Your Plate Campaign." Presented at the International Communication Association conference.

———. 2009a. "This Little Piggy Went to Press: The American News Media's Construction of Animals in Agriculture." *Communication Review* 12 (1) (March): 78–103.

———. 2009b. "A Greater Means to the Greater Good: Ethical Guidelines to Meet Social Movement Organization Advocacy Challenges." *Journal of Mass Media Ethics* 24 (4) (October): 269–288.

———. 2010a. "Embracing Humanimality: Deconstructing the Human/animal Dichotomy." In *Arguments About Animal Ethics*, edited by Greg Goodale and Jason Edward Black, 11–30. Lanham, MD: Lexington Books.

———. 2010b. "Framing Animal Rights in the 'Go Veg' Campaigns of U.S. Animal Rights Organizations." *Society & Animals* 18 (2): 163–182.

———. 2010c. "Meat's Place on the Campaign Menu: How U.S. Environmental Discourse Negotiates Vegetarianism." *Environmental Communication* 4 (3): 255–276.

———. 2013. "Stepping Up to the Veggie Plate: Framing Veganism as Living Your Values." In *Perspectives on Human-Animal Communication: Internatural Communication*, edited by Emily Plec. New York: Routledge.

Freeman, Carrie P., Marc Bekoff, and Sarah M. Bexell. 2011. "Giving Voice to the 'Voiceless:' Incorporating Nonhuman Animal Perspectives as Journalistic Sources." *Journalism Studies* 12 (5): 590–607.

Freston, Kathy. 2011. *Veganist: Lose Weight, Get Healthy, Change the World*. New York: Weinstein Books.

Gamson, William A. 1992. *Talking Politics*. Cambridge: Cambridge University Press.

Garner, Robert. 1993. *Animals, Politics, and Morality*. New York: St. Martin's Press.

Gitlin, Todd. 2003. *The Whole World is Watching: Mass Media in the Making & Unmaking of the New Left*. Berkeley: University of California Press.

Glenn, Cathy B. 2004. "Constructing Consumables and Consent: A Critical Analysis of Factory Farm Industry Discourse." *Journal of Communication Inquiry* 28 (1): 63–81.

Goffman, Erving. 1974. *Frame Analysis*. Harmondsworth: Penguin Books.

Hall, Lee. 2006a. *Capers in the Churchyard: Animal Rights Advocacy in the Age of Terror*. Darien, CT: Nectar Bat Press.

———. 2006b. "Animal Rights and Wrongs." *Satya*, October.

Hall, Stuart. 1975. "Introduction." In *Paper Voices: The Popular Press and Social Change, 1935-1965*, by Anthony Charles Smith, 11–24. Totowa, NJ: Rowman and Littlefield.

———. 1980. "Encoding/Decoding." In *Culture, Media, Language: Working Papers in Cultural Studies, 1972-79*, edited by Stuart Hall, Dorothy Hobson, Andrew Lowe, and Paul Willis, 128–138. London: Centre for Contemporary Cultural Studies, University of Birmingham.

———. 1982. "The Rediscovery of Ideology." In *Culture, Society, and the Media*, edited by Michael Gurevitch, Tony Bennett, James Curran, and Janet Woollacott, 56–90. London; New York: Methuen.

———. 1997. *Representation: Cultural Representations and Signifying Practice*. London: Sage.

Howard, Allen M. 2006. "Nineteenth-Century Coastal Slave Trading and the British Abolition Campaign in Sierra Leone." *Slavery & Abolition* 27 (1): 23–49.

Hoy, David Couzens. 1981. "Power, Repression, Progress: Foucault, Lukes, and the Frankfurt School." *TriQuarterly* 52: 43–63.

HRC. 2007. "Advocating Meat Reduction and Vegetarianism to U.S. Adults". Humane Research Council. http://www.humanespot.org/node/1956.

Humane Research Council. 2011. "Farmed Animal Indicators: HumaneTrends.com." August 15. http://www.humanetrends.org/farmed-animal-indicators/.

———. 2012. "Animal Tracker Survey - Year 5." May 1. http://www.humanespot.org/content/animal-tracker-year-5.

Humane Society International. 2011. "The Impact of Industrial Farm Animal Production on Food Security in the Developing World." Washington, D.C.: Humane Society International.

Irvin, David. 2007. "Control Debate, Growers Advised." *Arkansas Democrat Gazette*, September 22. http://www.nwanews.com/adg/Business/202171.

Jamieson, Kathleen Hall. 1988. *Eloquence in an Electronic Age*. New York: Oxford University Press.

Jasper, James. 1997. *The Art of Moral Protest: Culture, Biography, and Creativity in Social Movements*. Chicago: University of Chicago Press.

Jasper, James, and Dorothy Nelkin. 1992. *The Animal Rights Crusade: The Growth of a Moral Protest*. New York: Free Press.

Jepson, Jill. 2008. "A Linguistic Analysis of Discourse on the Killing of Nonhuman Animals." *Society & Animals* 16 (2) (June): 127.

Johnson, Victoria. 1997. "Operation Rescue, Vocabularies of Motive and Tactical Action: A Study of Movement Framing in the Practice of Quasi-nonviolence." *Research in Social Movements, Conflict and Change* 20: 103–150.

Jowett, Garth, and Victoria O'Donnell. 1999. *Propaganda and Persuasion*. 3rd ed. Thousand Oaks, CA: Sage Publications.

Joy, Melanie. 2010. *Why We Love Dogs, Eat Pigs, and Wear Cows: An Introduction to Carnism*. San Francisco: Conari Press.

———. 2011. "Understanding Neocarnism: How Vegan Advocates Can Appreciate and Respond to 'Happy Meat,' Locavorism, and 'Paleo Dieting'." July 29. http://www.onegreenplanet.org/lifestyle/understanding-neocarnism/.

Kim, Claire Jean. 2011. "Moral Extensionism or Racist Exploitation? The Use of Holocaust and Slavery Analogies in the Animal Liberation Movement." *New Political Science* 33 (3) (September): 311–333.

Lakoff, George. 2004. *Don't Think of an Elephant!: Know Your Values and Frame the Debate.* White River Junction, VT: Chelsea Green Pub. Co.

Lama, Eddie. 2006. "Sadly, Happy Meat." *Satya*, September.

Larsen, Janet. 2012. "Peak Meat: U.S. Meat Consumption Falling." Earth Policy Institute. http://www.earth-policy.org/data_highlights/2012/highlights25.

LaVeck, James. 2006a. "Compassion for Sale?" *Satya*, September.

———. 2006b. "Invasion of the Movement Snatchers." *Satya*, Oct.

Lawrence, Elizabeth Atwood. 1995. "Cultural Perceptions of Differences Between People and Animals: A Key to Understanding Human-animal Relationships." *Journal of American Culture* 18 (3): 75–82.

Leopold, Aldo. 2003. "The Land Ethic." In *The Environmental Ethics and Policy Book: Philosophy, Ecology, Economics*, edited by Donald VandeVeer and Christine Pierce, 3rd ed. Belmont, CA: Wadsworth Pub. Co.

Linzey, Andrew, and Paul Clarke. 2004. *Animal Rights: A Historical Anthology*. New York: Columbia University Press.

Loughnan, Steve, Nick Haslam, and Brock Bastian. 2010. "The Role of Meat Consumption in the Denial of Moral Status and Mind to Meat Animals." *Appetite* 55 (1): 156–159.

Lyman, Howard. 2006. "Straight Talk from a Former Cattleman." *Satya*, September.

Mark, Patty. 2006. "The Importance of Being Honest." *Satya*, Sept.

Martinson, David. 1996. "'Truthfulness' in Communication is Both a Reasonable and Achievable Goal for Public Relations Practitioners." *Public Relations Quarterly* 41 (4): 42–45.

Mason, Jim. 1997. *An Unnatural Order: Why We are Destroying the Planet and Each Other*. New York: Continuum.

Matheny, Gaverick. 2003. "Least Harm: A Defense of Vegetarianism from Steven Davis's Omnivorous Proposal." *Journal of Agricultural and Environmental Ethics* 16 (5): 505–511.

Maurer, Donna. 1995. "Meat as a Social Problem: Rhetorical Strategies in the Contemporary Vegetarian Literature." In *Eating Agendas: Food and Nutrition as Social Problems*, edited by Donna Maurer and Jeffery Sobal. New York: Aldine de Gruyter.

———. 2002. *Vegetarianism: Movement or Moment?* Philadelphia, PA: Temple University Press.

McAdam, Doug. 1996. "Framing Functions of Movement Tactics." In *Comparative Perspectives on Social Movements: Political Opportunities, Mobilizing Structures, and Cultural Framings*, edited by Doug McAdam, John D. McCarthy, and Mayer Zald, 338–356. Cambridge: Cambridge University Press.

McAdam, Doug, John D. McCarthy, and Mayer Zald. 1996. *Comparative Perspectives on Social Movements: Political Opportunities, Mobilizing Structures, and Cultural Framings*. Cambridge: Cambridge University Press.

McCance, Dawne. 2013. *Critical Animal Studies: An Introduction*. Albany, NY: State University of New York Press.

McChesney, Robert W. 1999. *Rich Media, Poor Democracy: Communication Politics in Dubious Times*. Urbana, IL: University of Illinois Press.

McGee, Michael Calvin. 1980. "The Ideograph: A Link Between Rhetoric and Ideology." *Quarterly Journal of Speech* 66 (1): 1–16.

McWilliams, James. 2009. *Just Food: Where Locavores Get it Wrong and How We Can Truly Eat Responsibly*. New York: Little, Brown and Company.

Melina, Vesanto, Brenda Davis, and Victoria Harrison. 1995. *Becoming Vegetarian: The Complete Guide to Adopting a Healthy Vegetarian Diet*. Summertown, TN: Book Publishing Co.

Midgley, Mary. 1984. *Animals and Why They Matter*. Athens, GA: University of Georgia Press.

———. 2008. "Why Farm Animals Matter." In *The Future of Animal Farming*, edited by Marian Stamp Dawkins and Roland Bonney, 21–31. Oxford: Blackwell Pub.

Mika, Marie. 2006. "Framing the Issue: Religion, Secular Ethics and the Case of Animal Rights Mobilization." *Social Forces* 85 (2): 915–941.

Munro, Lyle. 1999. "Contesting Moral Capital in Campaigns Against Animal Liberation." *Society and Animals* 7 (1): 35–53.

Newport, Frank. 2012. "In U.S., 5% Consider Themselves Vegetarians." Gallup. http://www.gallup.com/poll/156215/ConsiderThem selves-Vegetrians.aspx?utm_source=google&utm_medium= rss&utm_campaign=syndication.

Nibert, David Alan. 2013. *Animal Oppression and Human Violence: Domesecration, Capitalism, and Global Conflict*. New York: Columbia University Press.

Nierenberg, Danielle. 2003. "Factory Farming in the Developing World." *Worldwatch Institute*. http://www.worldwatch.org/node/ 534.

Norwood, Bailey, Jayson Lusk, and Robert Prickett. 2007. "Consumer Preferences for Farm Animal Welfare: Results of a Nationwide Telephone Survey." University of Oklahoma, Dept of Agricultural Economics. asp.okstate.edu/baileynorwood/Survey4/files/Initial ReporttoAFB.pdf.

Noske, Barbara. 1989. *Humans and Other Animals: Beyond the Boundaries of Anthropology*. London: Pluto Press.

Oliver, Pamela, and Hank Johnston. 2005. "What a Good Idea! Ideologies and Frames in Social Movement Research." In *Frames of Protest: Social Movements and the Framing Perspective*, edited by Hank Johnston and John Noakes, 185–203. Lanham, MD: Rowman & Littlefield Publishers.
Pace, Lesli. 2005. "Image Events and PETA's Anti Fur Campaign." *Women & Language* 28 (2): 33–41.
Park, Miyun. 2006. "Calculating Compassion." *Satya*, October.
Perelman, Chaïm, and Lucie Olbrechts-Tyteca. 1969. *The New Rhetoric: A Treatise on Argumentation*. Notre Dame, IN: University of Notre Dame Press.
PETA Annual Review. 2012. "PETA 2011: Annual Review." http://features.peta.org/Annual-Review-2011/.
PETA History. 2012. "PETA's History: Compassion in Action." http://www.peta.org/about/learn-about-peta/history.aspx.
PETA Mission. 2012. "About PETA: Our Mission Statement." http://www.peta.org/about/default.aspx.
Pew Environmental Group. 2007. "Protecting Life in the Sea". Philadelphia, PA: Pew Charitable Trust. http://www.pewtrusts.org/uploadedFiles/wwwpewtrustsorg/TaxonomyCopy/Enviroment/oceans_final_web.pdf.
Pollan, Michael. 2006. *The Omnivore's Dilemma: A Natural History of Four Meals*. New York: Penguin Press.
Polletta, Francesca. 2006. *It Was Like a Fever: Storytelling in Protest and Politics*. Chicago: University of Chicago Press.
Potter, Will. 2011. *Green Is the New Red : An Insider's Account of a Social Movement Under Siege*. San Francisco: City Lights Books.
Regan, Tom. 1975. "The Moral Basis of Vegetarianism." *Canadian Journal of Philosophy* 5 (2) (October): 181–214.
———. 1983. *The Case for Animal Rights*. Berkeley: University of California Press.
———. 2002. "How to Worry About Endangered Species." In *Environmental Ethics: What Really Matters, What Really Works*, edited by David Schmidtz and Elizabeth Willott, 105–108. New York: Oxford University Press.
———. 2003. *Animal Rights, Human Wrongs: An Introduction to Moral Philosophy*. Lanham, MD: Rowman & Littlefield Publishers.

———. 2004. "The Struggle for Animal Rights." In *Animal Rights: A Historical Anthology*, edited by Andrew Linzey and Paul Clarke, 176–186. New York: Columbia University Press.
Rifkin, Jeremy. 1992. *Beyond Beef: The Rise and Fall of the Cattle Culture*. New York: Dutton.
Robbins, John, and Jia Patton. 1992. *May All Be Fed: Diet for a New World*. New York: W. Morrow.
Rokeach, Milton. 1973. *The Nature of Human Values*. New York: Free Press.
Rolston, Holmes. 1993. "Challenges in Environmental Ethics." In *Environmental Philosophy: From Animal Rights to Radical Ecology*, edited by Michael E. Zimmerman, 135–157. Englewood Cliffs, NJ: Prentice-Hall.
Ryan, Charlotte. 1991. *Prime Time Activism: Media Strategies for Grassroots Organizing*. Boston, MA: South End Press.
Sagoff, Mark. 1993. "Animal Liberation and Environmental Ethics: Bad Marriage, Quick Divorce." In *Environmental Philosophy: From Animal Rights to Radical Ecology*, edited by Michael E. Zimmerman, 84–94. Englewood Cliffs, NJ: Prentice-Hall.
Salt, Henry Stephens. 1921. *Seventy Years Among Savages*. London: Allen & Unwin.
Schmidtz. 2002. "Are All Species Equal?" In *Environmental Ethics: What Really Matters, What Really Works*, edited by David Schmidtz and Elizabeth Willott, 96–103. New York: Oxford University Press.
Scully, Matthew. 2002. *Dominion: The Power of Man, the Suffering of Animals, and the Call to Mercy*. New York: St. Martin's Press.
Selman, Matt. 2000. "Lisa the Treehugger." *The Simpsons*. Fox.
Shove, Elizabeth. 2010. "Beyond the ABC: Climate Change Policy and Theories of Social Change." *Environment and Planning* 42 (6) (June): 1273–1285.
Simonson, Peter. 2001. "Social Noise and Segmented Rhythms: News, Entertainment, and Celebrity in the Crusade for Animal Rights." *The Communication Review* 4 (3): 399–420.
Singer, Peter. 1990. *Animal Liberation*. 2nd ed. New York: Random House.
———. 2006. "Singer Says." *Satya*, October.

———. 2008. "Foreword." In *The Future of Animal Farming*, edited by Marian Stamp Dawkins and Roland Bonney, vii–ix. Malden, MA: Blackwell Pub.

Singer, Peter, and Bruce Friedrich. 2006. "The Longest Journey Begins with a Single Step: Promoting Animal Rights by Promoting Reform." *Satya*, September.

Singer, Peter, and Jim Mason. 2006. *The Ethics of What We Eat: Why Our Food Choices Matter*. New York: Rodale.

Snow, David A, and Robert D. Benford. 1988. "Ideology, Frame Resonance, and Participant Mobilization." *International Social Movement Research* 1 (1): 197–217.

———. 2005. "Clarifying the Relationship Between Framing and Ideology." In *Frames of Protest: Social Movements and the Framing Perspective*, edited by Hank Johnston and John Noakes, 205–211. Lanham, MD: Rowman & Littlefield Publishers.

Snow, David A., E. Burke Rochford, Steven K. Worden, and Robert D. Benford. 1986. "Frame Alignment Processes, Micromobilization, and Movement Participation." *American Sociological Review* 51 (4): 464–481.

Spiegel, Marjorie. 1996. *The Dreaded Comparison: Human and Animal Slavery*. New York: Mirror Books.

Stahler, Charles. 2011. "How Often Do Americans Eat Vegetarian Meals?" The Vegetarian Resource Group. http://www.vrg.org/journal/vj2011issue4/vj2011issue4poll.php.

Stanescu, Vasile. 2010. "'Green' Eggs and Ham? The Myth of Sustainable Meat and the Danger of the Local." *Journal for Critical Animal Studies* 8 (1 and 2): 8–32.

Steiner, Gary. 2008. *Animals and the Moral Community: Mental Life, Moral Status, and Kinship*. New York: Columbia University Press.

Stewart, Charles J., Craig Allen Smith, and Robert E. Denton. 1984. "The Persuasive Function of Social Movements." In *Persuasion and Social Movements*, 51 – 82. Prospect Heights, IL: Waveland Press.

Tarrow, Sidney G. 1998. *Power in Movement: Social Movements and Contentious Politics*. New York: Cambridge University Press.

Taylor, Paul. 1993. "The Ethics of Respect for Nature." In *Environmental Philosophy: From Animal Rights to Radical Ecology*, edited by Michael E Zimmerman, 67–81. Englewood Cliffs, NJ: Prentice-Hall.

Therborn, Göran. 1980. *The Ideology of Power and the Power of Ideology*. London: NLB.

Tonsor, Glynn, and Nicole Olynk. 2010. "U.S. Meat Demand: The Influence of Animal Welfare Media Coverage." Kansas State University. www.agmanager.info/livestock/marketing/animalwelfare/MF2951.pdf.

Torres, Bob. 2006. "The Odd Logic of Welfarism." *Satya*, September.

———. 2008. *Making a Killing: The Political Economy of Animal Rights*. Oakland, CA: AK Press.

Twine, Richard. 2010. *Animals as Biotechnology: Ethics, Sustainability and Critical Animal Studies*. London: Earthscan.

Ucko, Peter J. 1988. "Foreword." In *What Is an Animal?*, edited by Tim Ingold. London: Unwin Hyman.

USDA. 2012. "U.S. Department of Agriculture Food Groups." http://www.choosemyplate.gov/food-groups/.

USDA Dairy. 2012. "U.S. Department of Agriculture Tips for Making Wise Choices in the Dairy Group." http://www.choosemyplate.gov/food-groups/dairy-tips.html#nomilk.

USDA Vegetarian. 2012. "U.S. Department of Agriculture Healthy Eating Tips for Vegetarians." http://www.choosemyplate.gov/healthy-eating-tips/tips-for-vegetarian.html.

Varner, Gary E. 1998. *In Nature's Interests? Interests, Animal Rights, and Environmental Ethics*. New York: Oxford University Press.

Vegan Outreach Financial. 2012. "Guidestar: Nonprofit Report." http://www.guidestar.org/organizations/86-0736818/vegan-outreach.aspx.

Vegan Outreach History. 2012. "A History of VO and Our Influences." http://www.veganoutreach.org/articles/history.html.

Vegan Outreach Leafleting. 2012. "Statistics." http://www.adoptacollege.org/totals.

Walters, Kerry S., and Lisa Portmess. 1999. *Ethical Vegetarianism: From Pythagoras to Peter Singer*. Albany, NY: State University of New York Press.

Webb, Janette. 2012. "Climate Change and Society: The Chimera of Behaviour Change Technologies." *Sociology* 46 (1): 109–125.

Wikipedia. 2012. "Vegetarianism by Country." http://en.wikipedia.org/wiki/Vegetarianism_by_country.

Wolfe, Cary. 2003. *Animal Rites: American Culture, the Discourse of Species, and Posthumanist Theory*. Chicago: University of Chicago Press.

Wood, David. 2004. "Thinking with Cats (a Response to Derrida)." In *Animal Philosophy: Essential Readings in Continental Thought*, edited by Matthew Calarco and Peter Atterton, 128–144. London: Continuum.

Worldwatch Institute. 2004. "Is Meat Sustainable?" http://www.worldwatch.org/node/549.

Younge, Gary. 2006. "'We're Stunt Queens. We Have to Be'" *The Guardian*. February 24. http://www.guardian.co.uk/uk/2006/feb/24/animalwelfare.comment.

Yourofsky, Gary. 2010. "Best Speech You Will Ever Hear - Gary Yourofsky - YouTube" Georgia Institute of Technology. http://www.youtube.com/watch?v=es6U00LMmC4.

Zald, Mayer. 1996. "Culture, Ideology, and Strategic Framing." In *Comparative Perspectives on Social Movements: Political Opportunities, Mobilizing Structures, and Cultural Framings*, edited by Doug McAdam, John D. McCarthy, and Mayer Zald, Chapter 11. New York: Cambridge University Press.

Zogby, Joseph. 2003. "Nationwide Views on the Treatment of Farm Animals." Animal Welfare Trust. http://civileats.com/wp-content/uploads/2009/09/AWT-final-poll-report-10-22.pdf.

Index

If you have any questions regarding this title, please contact:

Koninklijke Brill BV
Plantijnstraat 2
2321 JC Leiden
Email: info@brill.com

Batch number: 10723294